# Battery Safety Symposium 2016

AA006194

Held at the 33rd Annual International Battery Seminar & Exhibit 2016

Fort Lauderdale, Florida, USA
24 March 2016

ISBN: 978-1-5108-2254-2

**Printed from e-media with permission by:**

Curran Associates, Inc.
57 Morehouse Lane
Red Hook, NY 12571

CURRAN ASSOCIATES INC. proceedings.com

**Some format issues inherent in the e-media version may also appear in this print version.**



Printed by Curran Associates, Inc. (2016)

For permission requests, please contact International Battery Seminar
at the address below.

International Battery Seminar
250 First Avenue, Suite 300
Needham, MA 02494
USA

Phone: (561) 367-0193
Fax: (561) 367-8429

info@powersources.net

**Additional copies of this publication are available from:**

Curran Associates, Inc.
57 Morehouse Lane
Red Hook, NY 12571 USA
Phone: 845-758-0400
Fax: 845-758-2633
Email: curran@proceedings.com
Web: www.proceedings.com

SYMPOSIUM 1: MARCH 24, 2016

# BATTERY SAFETY

*Innovations to Improve Lithium Battery Safety from Cell to Systems*

**7:30 am** **Symposium Registration and Morning Coffee**

## CELL SAFETY, PERFORMANCE AND LIFETIME MANAGEMENT

**8:00** **Organizer's Opening Remarks....N/A**
*Mary Ann Brown, Executive Director, Conferences, Cambridge EnerTech*

**8:05** **Chairperson's Opening Remarks....N/A**
*Judith Jeevarajan, Ph.D., Research Director, Electrochemical Safety, Underwriters Laboratories, Inc.*

**8:10** **FEATURED PRESENTATION: Li-Ion Battery Safety Technologies and Their Implementation....1**
*Brian Barnett, Ph.D., Vice President, TIAX/CAMX Power*

**8:40** **All Tests and Standards Are Not Created Equal – Even if They Have the Same Name!....22**
*Cindy Millsaps, President & CEO, Energy Assurance LLC*

**9:10** **Enhanced Battery Safety by Applying Embedded Photonic Sensors....34**
*Antonio Nedjalkov, Dipl.-Ing., Research Associate, Department for Fiber Optical Sensor Systems, Heinrich Hertz Institute, Fraunhofer-Gesellschaft*

**9:40** **Coffee Break with Poster Viewing**

**10:10** **Safe Lithium-Ion Battery Designs for Use, Transportation and Second Use....49**
*Judith Jeevarajan, Ph.D., Research Director, Electrochemical Safety, Underwriters Laboratories, Inc.*
*Sponsored by* UL

**10:40** **Single-Point Probes for Lithium-Ion Battery Fault Detection and Temperature Monitoring....66**
*Corey T. Love, Ph.D., Materials Research Engineer, Alternative Energy Section, Chemistry Division, U.S. Naval Research Laboratory*

**11:10** **Addressing the Safety Issues Related to Air Transportation of Lithium-Ion Batteries with Effective Engineered Thermal Management Solutions....77**
*Richard Clark, Strategic Business Development Specialist, Morgan Advanced Materials*
*Sponsored by* Morgan Advanced Materials

**11:40** **Enjoy Lunch on Your Own**

## MODELING AND ENGINEERING A SAFE SYSTEM FOR VARIOUS APPLICATIONS

**1:20 pm** **Chairperson's Remarks....N/A**
*Brian Barnett, Ph.D., Vice President, TIAX/CAMX Power*

**1:25** **NREL Multiphysics Modeling Tools for Designing Safer Li-Ion Batteries....N/A**
*Ahmad A. Pesaran, Ph.D., Manager, Energy Storage Group, Transportation and Hydrogen System Center, National Renewable Energy Laboratory (NREL)*

**1:55** **Li-Ion Battery Characterization across Space and Time: Imaging Microstructure in 2D, 3D and 4D....N/A**
*Jeff Gelb, Senior Applications Development Engineer, Carl Zeiss X-ray Microscopy*

**2:25** **Effects of Cell Reversal on Li-Ion Batteries....92**
*E. Joseph Nemanick, Ph.D., Senior Member, Technical Staff, Energy Technology Department, The Aerospace Corporation*

**2:55** **Refreshment Break with Poster Viewing**

**3:30** **Experiments and Simulations for Battery Safety....N/A**
*Sergiy Kalnaus, Ph.D., Research Scientist, Computational Engineering and Energy Sciences, Oak Ridge National Laboratory (ORNL)*

**4:00** **Battery Safety Requires More than Safe Batteries....105**
*Todd Vanyo, Senior Systems Engineer, Texas Instruments*

**4:30** **Mesoscale Implications in Lithium-Ion Battery Thermal Behavior....112**
*Chien-Fan Chen, Energy Storage Lead, Energy and Transport Sciences Laboratory, Mechanical Engineering, Texas A&M University*

**5:00** **Close of Symposium & Close of International Battery Seminar & Exhibit**

# BATTERY SAFETY

# SPEAKER LISTING


Dr. Brian M. Barnett
Vice President
TIAX/CAMX Power
35 Hartwell Ave.
Lexington, MA 02421

Chien-Fan Chen
Energy Storage Lead
Energy & Transport Sciences Lab & Mechanical Engineer
Texas A&M University
College Station, TX 77840

Richard Clark
Specialist
Strategic Business Development
Morgan Advanced Materials
4000 Westchase Blvd.
Ste. 170
Raleigh, NC 27607

Jeff Gelb
Senior Applications Development Engineer
Carl Zeiss Microscopy, Inc.
1 Zeiss Dr.
Thornwood, NY 10594

Dr. Judith Jeevarajan
Research Director, Electrochemical Safety
Electrochemical Safety
UL LLC
333 Pfingsten Rd.
Northbrook, IL 60062

Dr. Sergiy Kalnaus
Research Scientist
Computational Engineering & Energy Sciences
Oak Ridge National Lab
1 Bethel Valley Rd.
PO Box 2008
MS 6164
Oak Ridge, TN 37831

Dr. Corey T. Love
Materials Research Engineer
Alternative Energy & Chemistry
US Naval Research Lab
4555 Overlook Ave. SW
Code 6113
Washington, DC 20375

Cynthia Millsaps
President & CEO
Energy Assurance LLC
5202 Belle Wood Court
Ste. 106
Buford, GA 30518

Dr. E. Joseph Nemanick
Senior Member Tech Staff
Energy Technology
Aerospace Corp.
2310 E. El Segundo Blvd.
PO Box 92957
Los Angeles, CA 90009-2957

Dr. Ahmad A. Pesaran
Energy Storage Group Manager
Transportation & Hydrogen System Center
National Renewable Energy Lab
15013 Denver W. Pkwy.
Golden, CO 80401

Todd Vanyo
Senior Systems Engineer
Texas Instruments, Inc.
12500 TI Blvd.
PO Box 660199
MS 8709
Dallas, TX 75266

CAMX Power

# Li-ion Battery Safety Technologies and Their Implementation

C. McCoy, S. Sriramulu and Brian Barnett

**Presentation at 33rd Annual International Battery Seminar & Exhibit**

Fort Lauderdale, FL
March 24th 2016

brian@camxpower.com

**CAMX Power**
35 Hartwell Avenue
Lexington, MA
02421-3102

www.CAMXPOWER.com





## OUTLINE

- Let's watch thermal runaway
- Taking a closer look at internal shorts
- Short detection technology and its implementation
- If a cell needs to get to a certain temperature before ...
- Investigation of the nail penetration test and the "hard" short
- Summary

B.Barnett, D.Ofer, S.Sriramulu, R. Stringfellow (Safety Issues in Li-Ion Batteries) in: Robert A. Meyers (Ed.), Encyclopedia of Sustainability Science and Technology, Springer Science, 2012

**There are three broad categories of safety failures: each can lead to thermal runaway yet each exhibits very different physics.**

| Category of Safety Failure | Some Important Characteristics | Examples of Triggers |
|---|---|---|
| Simple Abuse<br>(Not discussed further today) | • Homogeneous Temperature Distribution<br>• Variable Stresses of T and t | Heat Exposure<br>Over Charge |
| | Discussed later in this presentation | |
| | Discussed later in this presentation | |

**Each category brings different implications for safety testers, safety modelers, regulators, first responders, battery developers, product developers.**





**What does a thermal runaway look like … if you were there?**

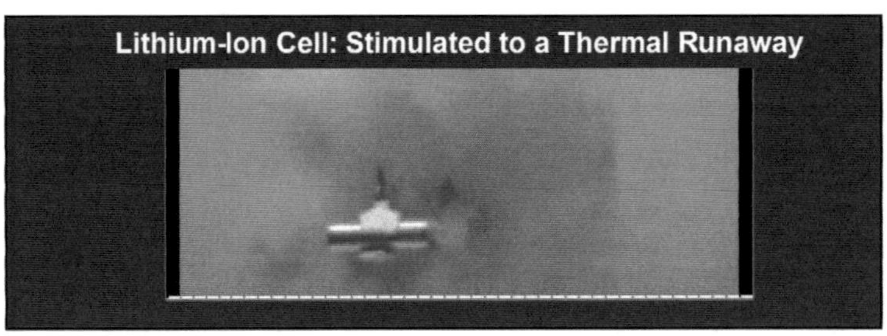


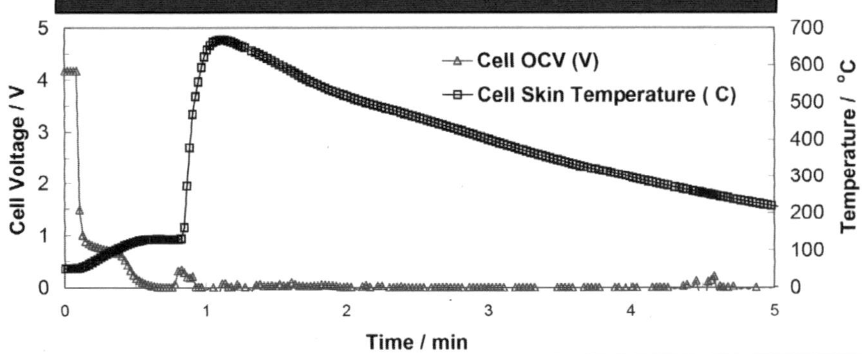

## Peak temperatures are very impressive – in this case, nearly 700°C.

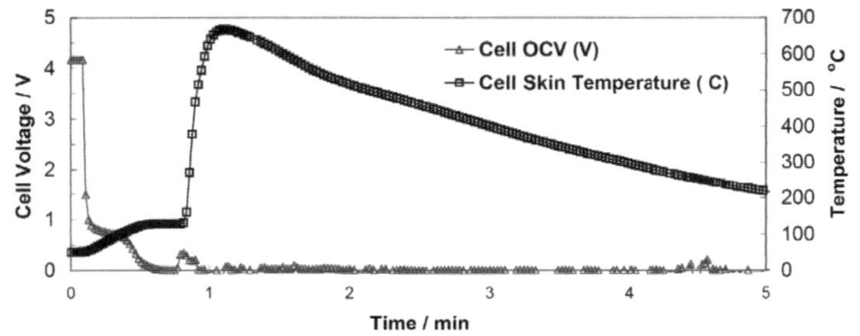


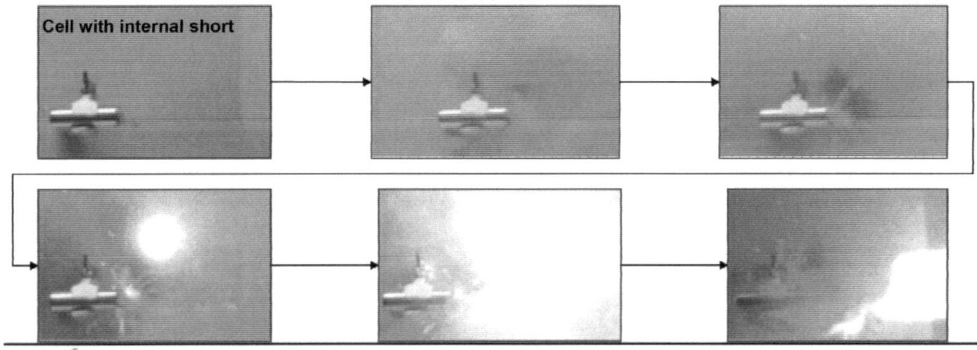






## A combination of non-invasive evaluation and post-mortem examination can help reconstruct the progression of the safety incident.

***Photos of post-mortem cells are removed for reasons of confidentiality.***

***Photos illustrate silver-colored beads characteristic of re-condensed aluminum (which melts at 660°C).***

Safety Perspective for Lithium-Ion Batteries

**The electrical energy in the cell alone is sufficient to raise the cell temperature over 700°C under adiabatic conditions.**

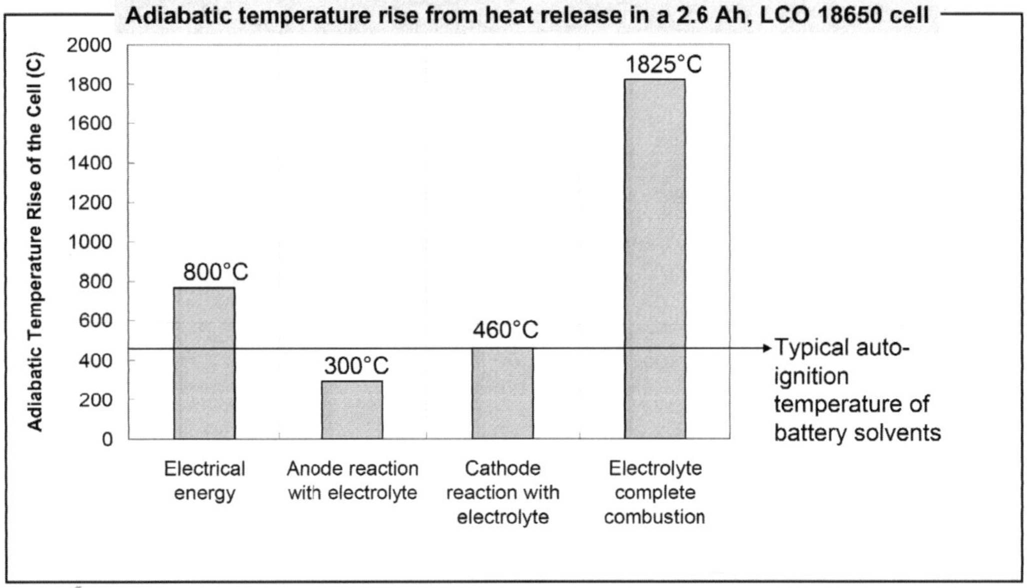




Safety Perspectives for Lithium-Ion Batteries

**Sometimes we hear ......**

***1. But these events aren't still happening, right ??***

***2. We couldn't reproduce them !***

## ...... but this week, during this conference, ...

### Recall date: MARCH 21, 2016

**Recall Summary**

**Name of product:**
Lithium-ion Computer Battery Packs

**Hazard:**
Conductive foreign material was mixed into the battery cells during manufacturing, posing a risk of fire.

**Remedy:**
View Details





**Taking a closer look at internal shorts ...**

## CAMX Power has augmented a commercial FEA program with subroutines for transient 3-d simulations of safety-related field-failures.

**Inputs:**

Heat generation associated with a short circuit:

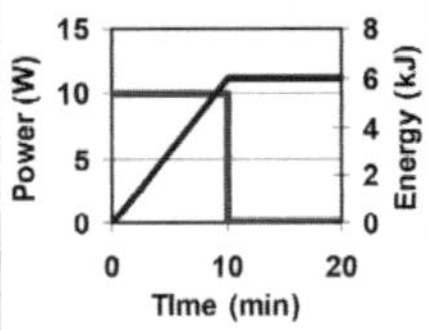


**Boundary conditions:**

- Surface heat transfer coefficient
- Ambient temperature

**Model properties:**

Kinetic models for thermal decomposition reactions in anode and cathode

Thermal properties for all cell constituents:
- density
- specific heat
- conductivity

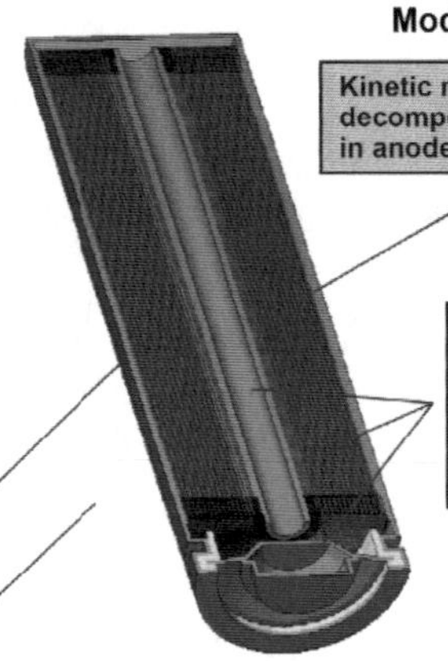

3-D transient FEA model with coupled decomposition rates & heat transfer

**Outputs:**

- Distribution of T, Qdot at selected time intervals.

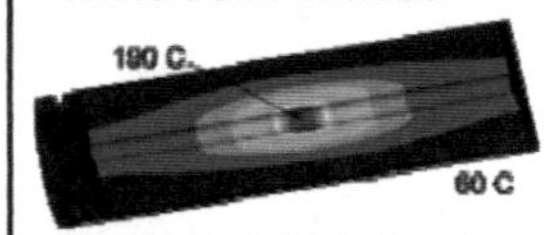


- Temperature histories at selected points in the cell.

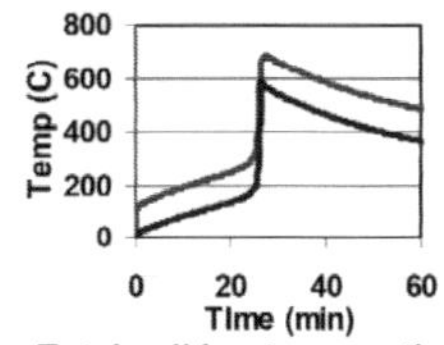


- Total cell heat generation rates, surface heat loss, etc.

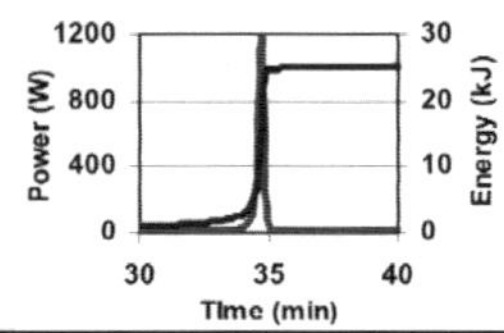






## Li-ion cells contain energetic materials, which are stimulated to release heat in a field-failure.

| Process | Temperature Range (°C) | Energy release[1] | Energy release in an 18650 cell |
|---|---|---|---|
| Decomposition at anode | 80 -120<br>150 - 300 | 300 – 450 J/ g-anode<br>1200 – 1400 J/g-anode | ~ 9.6 – 13.3 kJ |
| Decomposition at cathode | 150 - 300 | 1500 – 1800 J / g-cathode | ~ 16.8 – 30 kJ |
| Self-reaction of salt with solvent | 250 - 400 | 900 J/g electrolyte | ~2.5 – 5.4 kJ |
| Complete combustion of solvent[2] | Auto-ignition temperature ~ 450 | 18 kJ / g solvent | 90 – 126 kJ |

[1] Approximate values estimated from DSC and ARC testing of cell components: charged anodes and cathodes, and typical electrolyte compositions;

[2] Please note that there is insufficient oxygen available inside an 18650 cell to effect complete combustion of the solvent. However, if vented at high temperatures or vented in the presence of an ignition source the solvent can burn outside the cell.

## FEA simulations show the temperature evolution and heat release profiles when an internal short results in a thermal runaway.

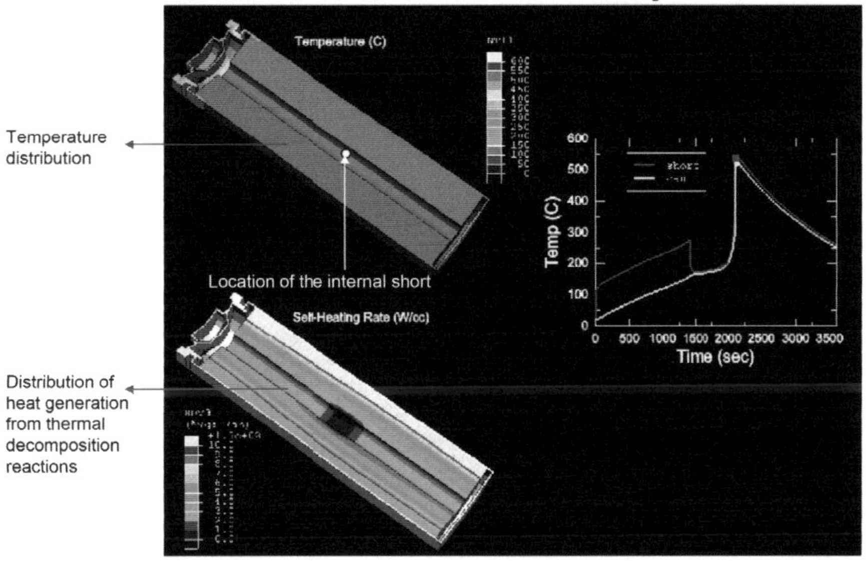


B.Barnett, D.Ofer, R.Stringfellow, S.Sriramulu (Safety Issues in Li-Ion Batteries) in: Robert A. Meyers (Ed.), Encyclopedia of Sustainability Science and Technology, Springer Science, 2012





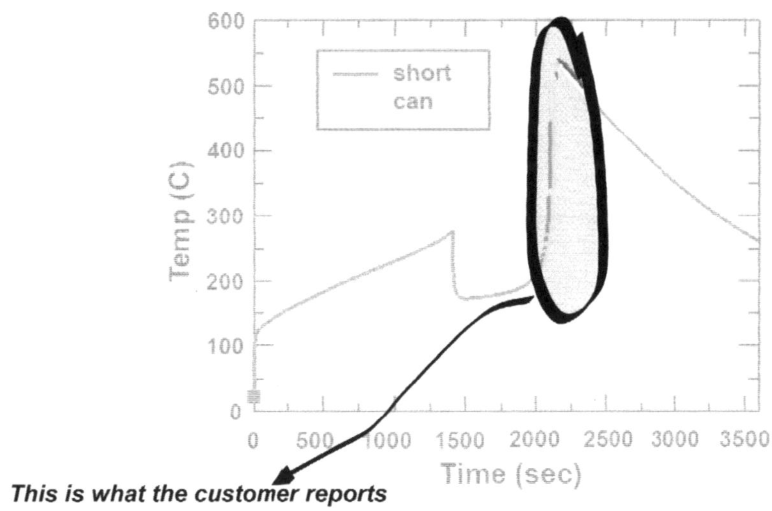

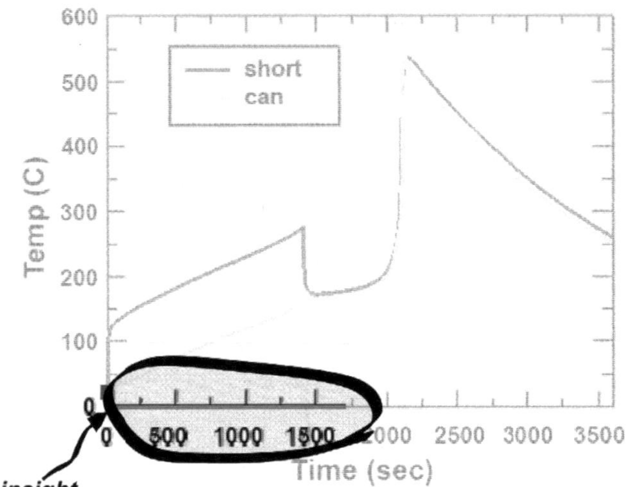


***This is new insight***

- Regardless of the input power, we appear to have several minutes before thermal runaway is inevitable, i.e., we exceed threshold power and then exceed threshold energy
- Intervention may be possible, but can be very difficult.





## *At what point does an internal short become a threat to cell safety?*

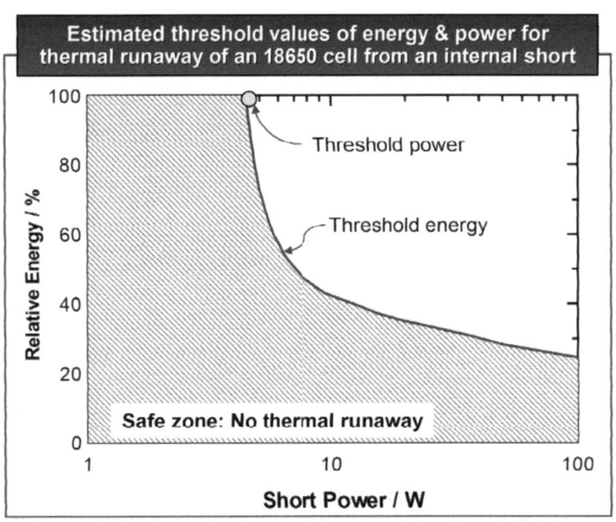


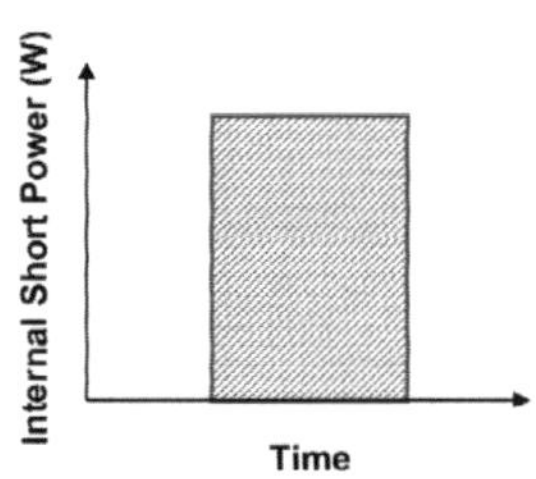


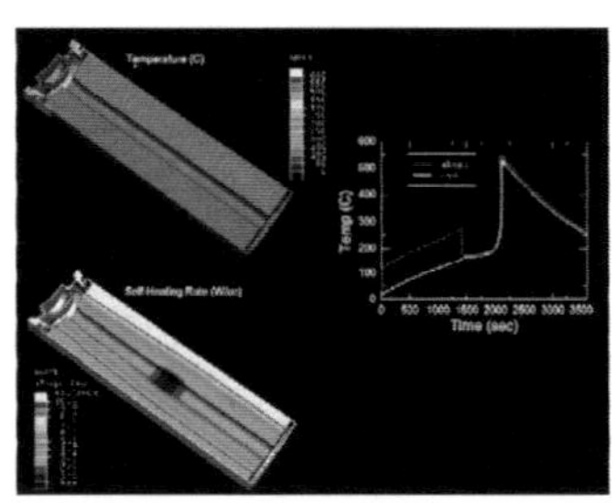

* For a 2.6 Ah 18650 cell with an external heat transfer coefficient of 10 W/m$^2$-K. The threshold values depend on the external heat transfer coefficient, cell design & chemistry.

## Implementation of short detection technology





## Thermal runaway in a Li-ion cell induced using unique particle implantation methodologies developed by CAMX Power.

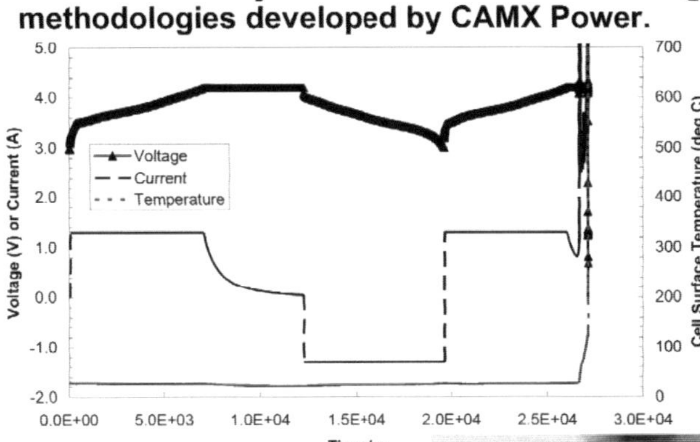

a) This cell ran away after a particle was implanted. Aluminum beads indicate cell temperatures exceeding 660°C.

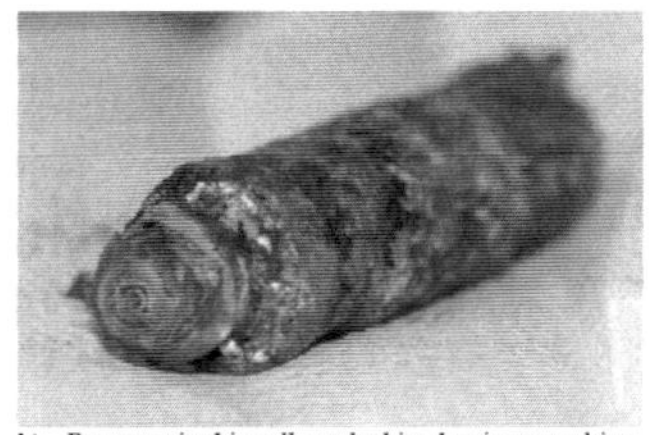

b) Runaway in this cell resulted in aluminum melting (beads can be seen on the end of the jellyroll) and partial ejection of the jellyroll core.

**Deliberate Metal Particle Implantation in 18650 Cells** Key results

## We successfully demonstrated that we can induce thermal runaway during normal operation of an 18650 cell (≥ 2.6 Ah).

This cell ran away after a particle was implanted. Aluminum beads indicate cell temperatures exceeding 660°C.

Runaway in this cell resulted in aluminum melting (beads can be seen on the end of the jellyroll) and partial ejection of the jellyroll core.

**We have succeeded in creating an internal short AND a thermal runaway after many charge-discharge cycles in an 18650 cell.**



**Internal Short Maturation**

## An internal short may develop over multiple cycles and/or over a long period of time, remaining virtually invisible to battery management for most of its life.

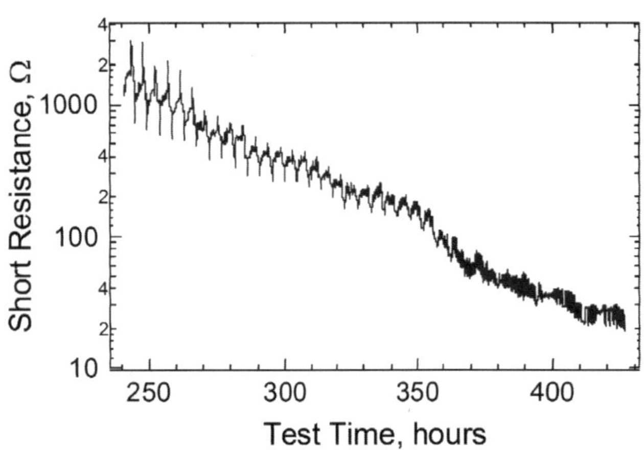


***Characterization of an internal short produced in a lithium-ion cell in CAMX Power labs by implantation of a metal particle***

**Analytically we determined a threshold power of ~ 4 W (a short resistance of ~4Ω) dissipated in an internal short or for thermal runaway of an 18650 cell**

- *Short resistance is initially >1000 ohms, giving rise to only a few milliamps of current – a level extraordinarily difficult to discern from the currents of normal charge and discharge in cells, which are three orders of magnitude larger.*

**The CAMX Power internal short circuit detection technology provides sensitive monitoring for internal shorts and can detect and quantify shorts (even minor shorts) in the 1000 ohm range.**

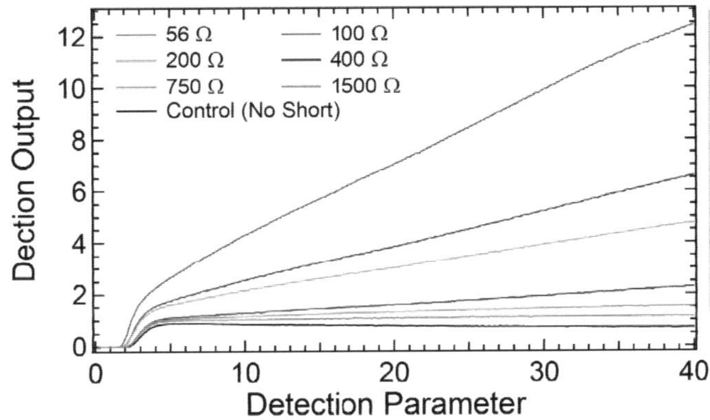

**The prototype on the right is part of a paid technology demonstration now underway at a major automotive manufacturer.**





**Automotive drive cycles pose one of the most challenging detection environments.**

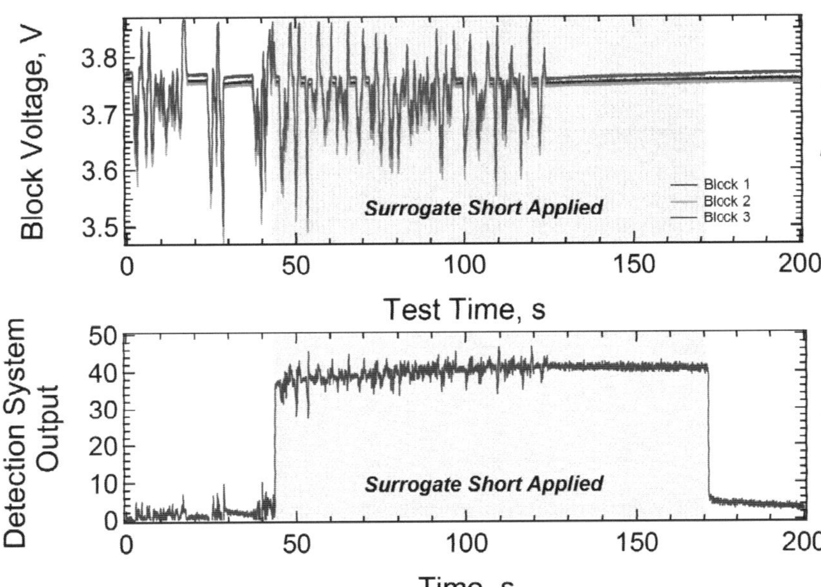


- *Standard BMS monitoring does not reveal the presence of a short*
- *CAMX Power detection technology does*

**There are three broad categories of safety failures: each can lead to thermal runaway yet each exhibits very different physics.**

| Category of Safety Failure | Some Important Characteristics | Examples of Triggers |
|---|---|---|
| Simple Abuse | • Homogeneous Temperature Distribution<br>• Variable Stresses of T and t | Heat Exposure<br>Over Charge |
| Grown-In Internal Short | • Non-Homogeneous Temp. Distribution<br>• Variable Stresses include short resistance, heat transfer, cell design<br>• "Long" but variable incubation period | Metal particle (aka FOD) seeds dendrites |
| | Discussed later in this presentation | |

**Each category brings different implications for safety testers, safety modelers, regulators, first responders, battery developers, product developers.**





## *Implementation of short detection technology*

CAMX Power Internal Short Detection Technology

## The CAMX Power Short Detection Technology is being adapted for a wide range of applications for enhanced safety and reduced costs.

**CAMX Power Internal Short Detection Technology**

- **Pack – field monitoring of cells**
  - PHEV battery trials
  - Cap-lamps worn by miners
  - Navy batteries
- **Pack Assembly – cell screening**
  - Underwater submersibles
- **Failure Audits - screening of cells/packs post-use**
  - Dreamliner battery investigation
- **Cell Factory – reduced aging time during formation**
  - Defense Logistics Agency

CAMX Power 24

Example: Short Detection for Mining Application

## One of the technologies is being adopted for detecting internal short circuits in cap lamps worn by miners (courtesy NIOSH*).

- Li-on based batteries (e.g., in cap lamps) are extensively employed in mining today.
- Internal short circuits in these batteries can pose a significant danger.
- Early warning of growing internal short circuits can enable intervention and prevent catastrophic/cascading failures.

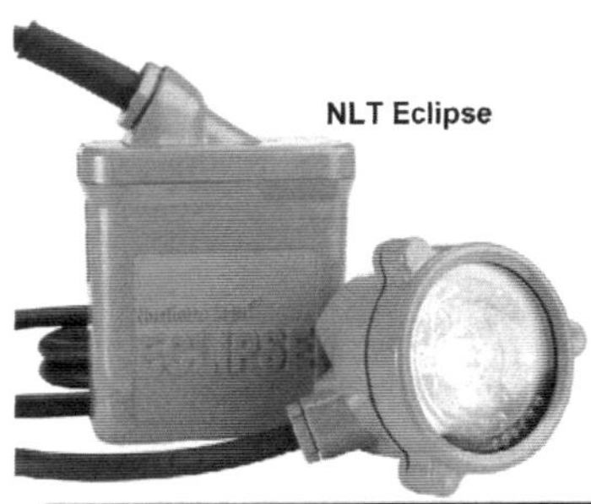


Initial prototype board with CAMX Power electronics to demonstrate internal short detection

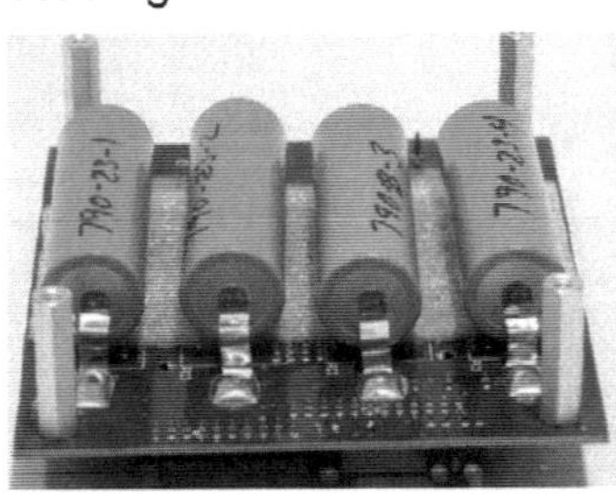

*Disclaimer: The findings and conclusions in this presentation have not been formerly disseminated by the National Institute for Occupational Safety and Health and should not be construed to represent any agency determination or policy. Mention of any commercial product does not imply endorsement.

CAMX Power 25

Example: Short Detection for Mining Application

**Our technology can provide real-time detection of internal short circuits in cells used in the miner's cap lamps.**

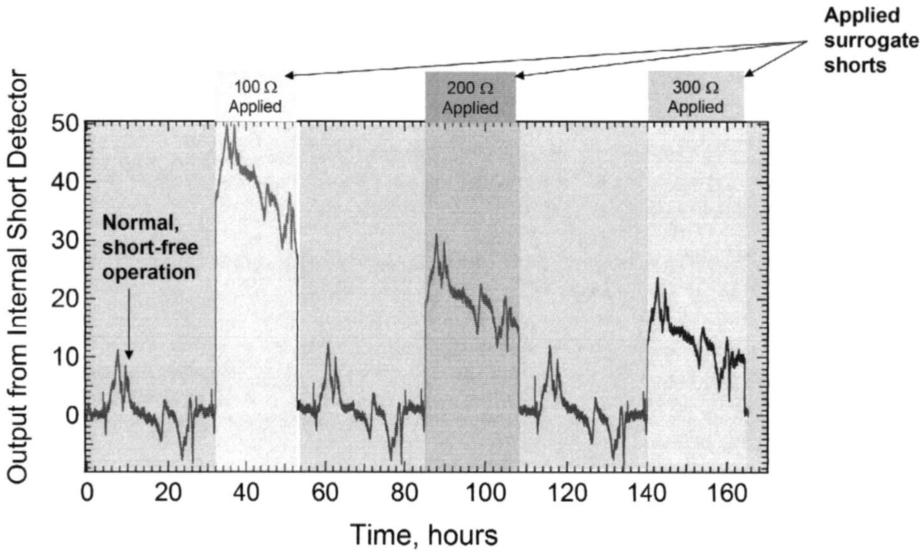


CAM Power

"The findings and conclusions in this presentation have not been formerly disseminated by the National Institute for Occupational Safety and Health and should not be construed to represent any agency determination or policy. Mention of any commercial product does not imply endorsement."

26

Example: Short Detection for Mining Application

**Our technology can provide real-time detection of internal short circuits in cells used in the miner's cap lamps.**

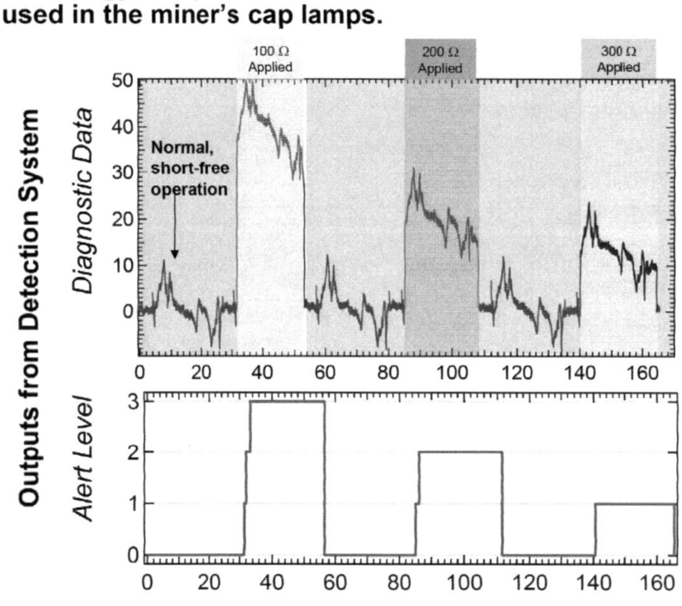


CAM Power

"The findings and conclusions in this presentation have not been formerly disseminated by the National Institute for Occupational Safety and Health and should not be construed to represent any agency determination or policy. Mention of any commercial product does not imply endorsement."

27

**Energy injection, timing to runaway**

**If a cell needs to get to a certain temperature before ...**





## Heating the cell to 100°C or 150°C takes a significantly long time under typical abuse conditions.

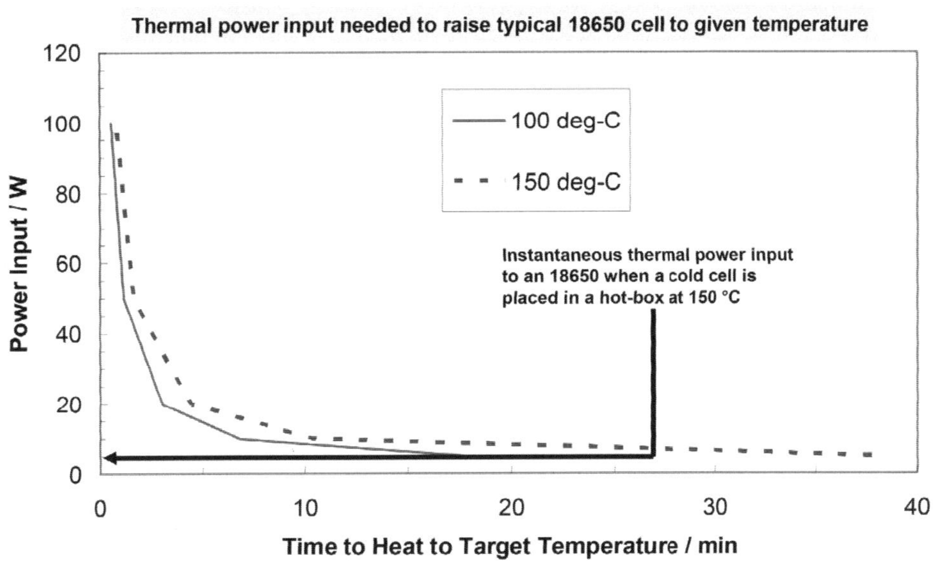

## Heating the cell to 130°C takes a significantly long time under typical abuse conditions.

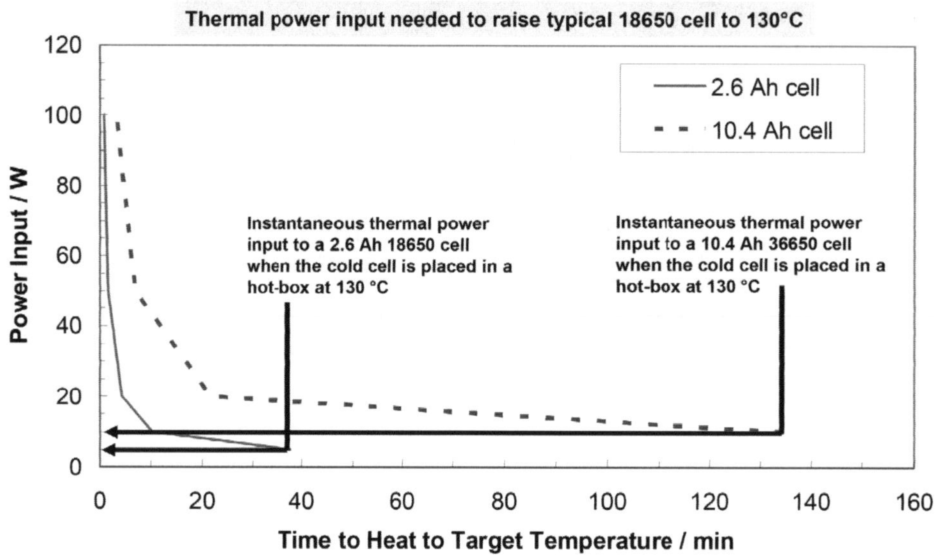






## Investigation of the Nail Penetration Test and the "Hard Short"

### Factors That Influence Thermal Runaway During a Nail Penetration Test

t = 126 ms after nail penetrated the can wall

t = 158 ms after nail penetrated the can wall

t = 190 ms after nail penetrated the can wall

**During the nail penetration test, it is not uncommon to observe explosions within 200 ms of nail penetration.**

**Our experimental setup allows us to characterize the nail penetration process in detail.**



Fast nail, 8 cm/s



### Li-ion Battery Safety   Insights from Nail Penetration Tests

## Analysis of the nail penetration data shows that the outcome in the nail penetration is governed by phenomena in the vicinity of the nail.

**During the nail penetration test, venting and explosions occur typically < 500 ms**

t = 126 ms after nail penetrated the can wall

t = 158 ms after nail penetrated the can wall

t = 190 ms after nail penetrated the can wall

**However, the cell surface temperature barely increases in this short time period.**

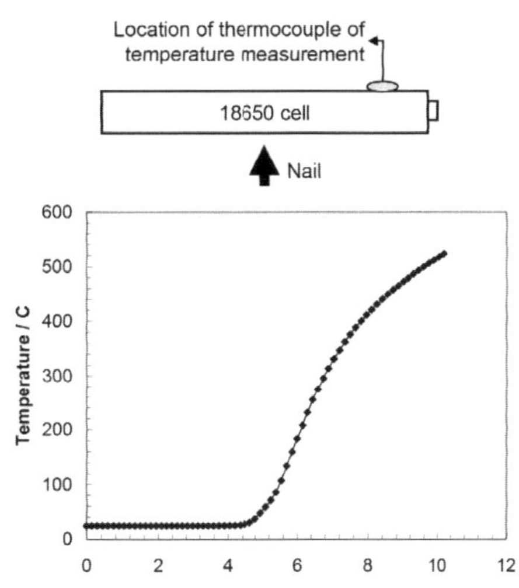

Nail Penetration Testing

## We undertook a detailed study of the hard-short situation in a nail penetration test using a combination of experiments and analysis.

High-speed photography of nail penetration

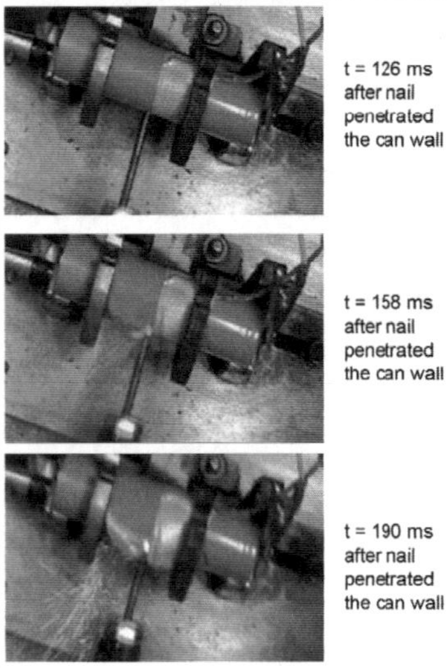


Simulation of Nail Penetration

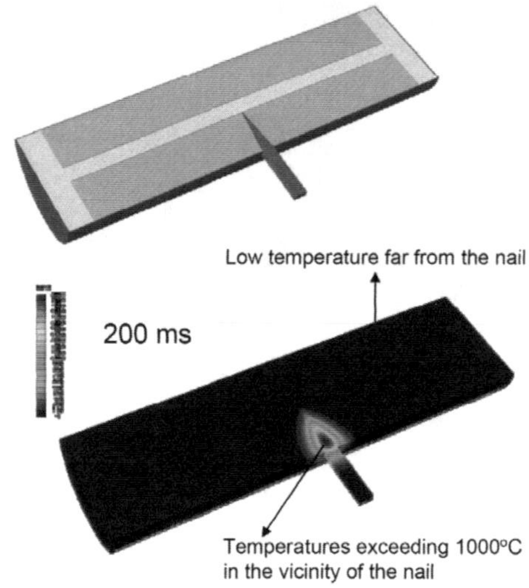


CAMX Power 34

Assuring Safety in Li-Ion Batteries: The hard short

## This cell shows violent thermal runaway in nail penetration tests.

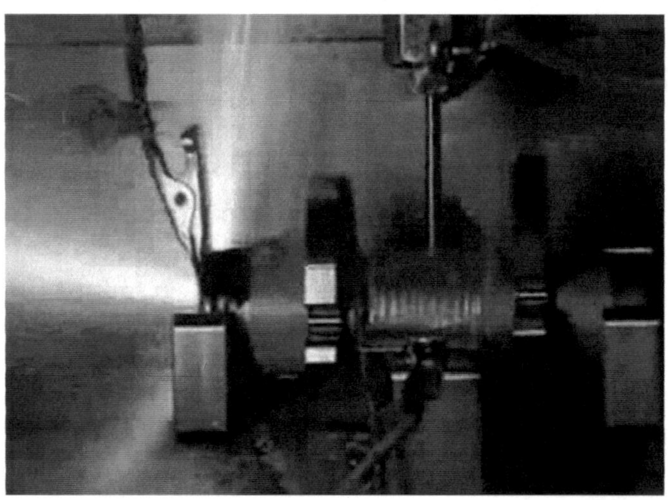

CAMX Power 35

## These cells with only modest modification but exactly the same anode and cathode show more benign response.

## SUMMARY

**There are three broad categories of safety failures: each can lead to thermal runaway yet each exhibits very different physics.**

| Category of Safety Failure | Some Important Characteristics | Examples of Triggers |
|---|---|---|
| Simple Abuse | • Homogeneous Temperature Distribution<br>• Variable Stresses of T and t | Heat Exposure<br>Over Charge |
| Grown-In Internal Short | • Non-Homogeneous Temp. Distribution<br>• Variable Stresses include short resistance, heat transfer, cell design<br>• "Long" but variable incubation period | Metal particle (aka FOD) seeds dendrites |
| Impact/Mechanical Intrusion | • Non-Homogeneous Temp. Distribution<br>• Variable Stresses include speed, angle, penetration, momentum | Impact<br>Crash<br>Crush |

**Each category brings different implications for safety testers, safety modelers, regulators, first responders, battery developers, product developers.**



Lithium-Ion Battery Safety Technologies and Their Implementation

**There are three broad categories of safety failures and we need somewhat different testing, modeling and design approaches for each category.**

| Category of Safety Failure | Some Important Characteristics | Examples of Triggers |
|---|---|---|
| Simple Abuse | • ANY Li-ion cell of a given cell design and using the same materials **WILL** exhibit the same safety "result" in response to the same trigger. | Heat Exposure<br>Over Charge |
| Grown-In Internal Short | • Very rare; less than one cell in 1-10 million<br>• Dependent on specific conditions of cell fabrication and subsequent field exposure | Metal particle (aka FOD) seeds dendrites |
| Impact/Mechanical Intrusion | • It would be desirable that ANY cell of a given cell design exhibit the "same" safety result in response to the same trigger.<br>• In practice, stochastic results are often observed. | Impact<br>Crash<br>Crush |

**A "grown-in" internal short-circuit begins from infinite/high resistance and matures to lower resistances, heating the cell, until the cell is "dead" or a thermal event occurs ...**

Depiction of the expected "steady state" power dissipation through an internal short circuit as a function of the short resistance and cell power capability

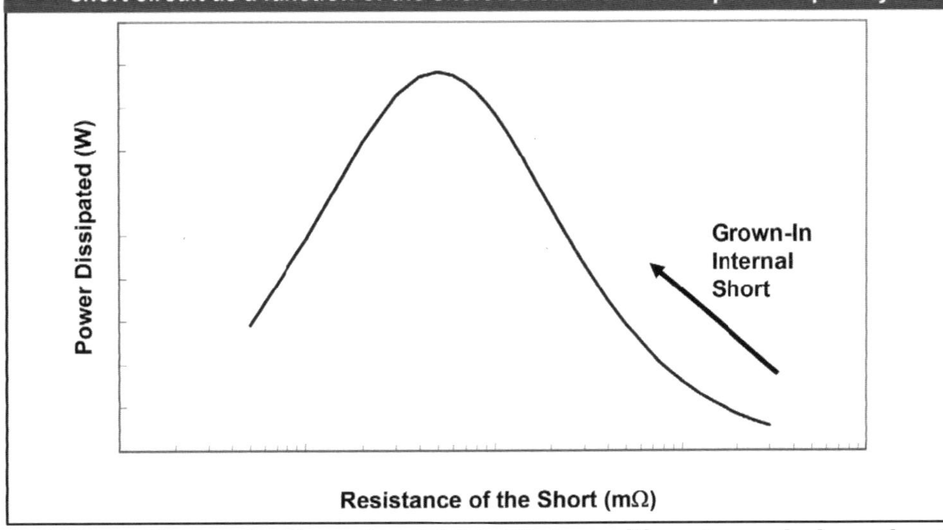


**... but the nail penetration short (s) initiates with extremely low short resistance, with little cell heating possible before an explosion occurs.**

ENERGY ASSURANCE

## All tests and standards are not created equal – even when they have the same name!

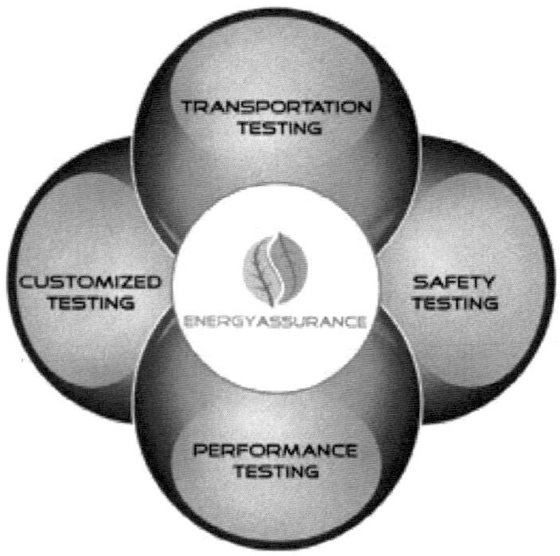


**Cynthia Millsaps**
**Energy Assurance, LLC**
**March 24, 2016**

*Performance - Compliance - Success*

ENERGY ASSURANCE

## What are we going to cover?

- Small
- Li Ion (including Lithium Polymer)/rechargeable
- Cells and batteries
- For use in portable devices/applications
- Covered by UN 38.3, UL 2054/1642 and IEC 62133

*Performance - Compliance - Success*

ENERGY ASSURANCE

## What are the current revisions of the standards and what do they cover

| Standard | Scope within our discussion |
| --- | --- |
| UN 38.3 6th revised edition | Li ion cells and batteries |
| UL 2054 ed 2, with revisions dated 09-2011 | Li ion batteries, including single cell batteries |
| UL 1642 ed 5, with revisions dates 06-2015 | Li ion cells only |
| IEC 62133:2012 | Li ion cells and batteries |

*Performance - Compliance - Success*

ENERGY ASSURANCE

## What questions do we commonly hear?

- The cells were already tested to the standard – why do the batteries have to be tested?
- There is an XXX test in both standards, why do you have to test it again?

*Performance - Compliance - Success*

ENERGY ASSURANCE

## What tests are part of UN38.3

| Cells/Single cell batteries | Batteries |
|---|---|
| Altitude Simulation (T1) | Altitude Simulation (T1) |
| Thermal Test (T2) | Thermal Test (T2) |
| Vibration (T3) | Vibration (T3) |
| Shock (T4) | Shock (T4) |
| External Short Circuit (T5) | External Short Circuit (T5) |
| Impact/Crush (T6) | Overcharge (T7) |
| Forced Discharge (T8) | |

Cell to battery overlap –
Addressed in the UN Manual of Tests for single and multiple cell applications

*Performance - Compliance - Success*

ENERGY ASSURANCE

## UL 2054/UL 1642

| Cells - UL 1642 | Battery - UL 2054 |
|---|---|
| Short Circuit at 55 | Short Circuit at 55 |
| Short Circuit RT | Short Circuit at RT |
| Abnormal Charge | Abnormal Charging |
| Forced Discharge | Abusive Overcharge |
| Crush | Forced Discharge |
| Temperature cycling | Drop Impact |
| Low pressure (Altitude Simulation) | |
| Projectile | |

Limited overlap can be addressed with reduced test plan in single cell applications only. Main difference – battery faults

*Performance - Compliance - Success*

ENERGY ASSURANCE

## IEC 62133

| Cells | Batteries |
|---|---|
| Continuous Charge | Moulded Case Stress |
| External Short Circuit | External Short Circuit |
| Free Fall | Free Fall |
| Thermal Abuse | Overcharge |
| Crush | |
| Forced Discharge | |
| Forced Internal Short Circuit | |

Cell and battery overlap –
Free fall may be able to be exempted
External Short test is at different values – no overlap exists – cell – RT, battery - 55

*Performance - Compliance - Success*

ENERGY ASSURANCE

## Summary of cell to battery overlap

Single cell battery:

- UN 38.3
  - T1-T5
- UL 1642-UL2054
  - Short Circuit at RT and 55
  - Abnormal Charge
- IEC 62133
  - Free fall – in some cases

Multiple cell battery:

- UN 38.3
  - No overlap
- UL 1642-UL 2054
  - No overlap
- IEC 62133
  - No overlap

*Performance - Compliance - Success*

ENERGY ASSURANCE

## Cell level testing across standards

| | UN 38.3 | UL 1642 | IEC 62133 |
|---|---|---|---|
| External Short Circuit | X | X | X |
| Forced Discharge | X | X | X |
| Crush/Impact | X | X | Crush only |
| Altitude Simulation/Low Pressure | X | X | N/A |
| Temperature Cycling | X | X | N/A |
| Vibration | X | X | N/A |
| Shock | X | X | N/A |
| Thermal Abuse/Heating | N/A | X | X |
| Abnormal Charge | N/A | X | N/A |
| Continuous Charge | [illegible] | [illegible] | X |
| Free Fall/drop | [illegible] | [illegible] | X |
| Projectile | N/A | X | N/A |

*Performance - Compliance - Success*

ENERGY ASSURANCE

## External Short Circuit – Cells

| | UN 38.3 | UL 1642 | IEC 62133:2012 |
|---|---|---|---|
| Samples under test | 10 fresh cells – fully charged | 5 fresh and 5 aged - fully charged | 5 fresh charged at low temp – 5 fresh charged at high temp – all fully charged |
| Test ambient | 57°C ± 4°C (6hr soak) – NEW in 6th edition | 20 and 55°C (±5°C) | 20°C (±5°C) |
| Circuit Resistance | <0.1Ω | 80±20mΩ | 80±20mΩ |
| Termination Criteria | 1hr or case returns to ambient | 0.2V or case returns to within 10°C of test ambient | 24h or until external temp declines by 20% from the maximum recorded temp rise |
| Pass/fail Criteria | Case temp <170°C, no disassembly, no rupture, no fire during the test and for 6 hours after termination | No fire/No explosion | No fire/No explosion |
| Other considerations | Test 5 in a sequence of tests | Protective devices? | |

*Performance - Compliance - Success*

## Forced Discharge – Cells

ENERGY ASSURANCE

| | UN 38.3 | UL 1642 | IEC 62133:2012 |
|---|---|---|---|
| Samples under test | 10 fresh and 10 cycled cells – all fully discharged | 5 fresh and 5 aged* - fully discharged<br>Plus enough cells to simulate intended series connection for each cell tested | Multiple cell application only<br>5 fresh FC cells |
| Test ambient | 20±5°C | 20 ±5°C | 20±5°C |
| Test procedure | Loaded for initial current = maximum discharge current | Connect one FD cell in series with the specified number of full charged cells and then short the string (80±20mΩ) | Reverse charge cell at $1I_tA$ |
| Termination Criteria | Time = rated capcity/initial test current | 0.2V or case returns to within 10°C of test ambient | Time = 90 minutes |
| Pass/fail Criteria | No disassembly and not fire during testing and for 7 days after termination | No fire/No explosion | No fire/No explosion |
| Other considerations | | Normally not done at cell level | |

*Performance - Compliance - Success*

## Crush/Impact – Cells

ENERGY ASSURANCE

| | UN 38.3 | UL 1642 | IEC 62133:2012 |
|---|---|---|---|
| Samples under test | Impact – cylindrical cells <18.0mm dia. Only<br>Crush for all others<br>5 fresh cells at 50% SOC | 5 fresh and 5 aged* - fully charged | Crush only – no impact requirement<br>5 fresh charged at low temp and 5 fresh charged at high temp<br>All FC cells |
| Test ambient | 20±5°C | 20 ±5°C | 20±5°C |
| Test procedure | Impact - 15.8mm dia. Rod, 6cm long; Type 316 stainless steel<br>9.1kg weigh dropped from 61cm<br><br>Crush – 13kN | Impact – 15.8mm dia. rod<br>9.1kg weight dropped from 61cm<br><br>Crush – 13kN | Impact – N/A<br><br>Crush – 13kN |
| Termination Criteria | Impact – single impact<br><br>Crush – Voltage drop 100mV; cell deforms by 50%; max pressure reached | Impact – Single impact<br><br>Crush – until force reached | Impact – N/A<br><br>Crush – abrupt voltage drop, or 10% deformation or max pressure reached |
| Pass/fail Criteria | 170°C, no disassembly, no fire during test and for 6hr after termination | No fire/No explosion | No fire/No explosion |
| Other considerations | Prismatic/pouch cells crushed along widest side<br>Button cells on flat surface | Prismatic/pouch cells crushed on widest side<br>Button cells on flat surface | Wide side crush only |

*Performance - Compliance - Success*

ENERGY ASSURANCE

## UN 38.3 T1-T4 to UL 1642 – Cells

| | UN 38.3 | UL 1642 |
|---|---|---|
| Altitude simulation (T1) / low pressure | 10 fresh fully charged cells<br>11.6kPa for at least 6hrs at 20±5°C | 5 fresh fully charged cells<br>11.6kPa for at least 6hrs at 20±3°C |
| Temperature Cycling (T2) | 10 fresh fully charged cells<br>Test 2 in sequential test program<br>+75±2°C and -40±2°C in 30 min transition and 6 hours at temp for 10 cycles then 24 wait | 5 fresh fully charged cells<br><br>+70±3°C, and 20±2°C and -40±2°C in 30 min transition and 4 hours at high and low temp, 2h at 20 for 10 cycles then 24 wait |
| Vibration (T3) | 10 fresh fully charged cells<br>Test 3 in sequential test program<br>Frequency sweep between 7Hz and 200Hz and back in 15 minutes for 12 cycles (3 hours per plane) | 5 fresh fully charged cells<br><br>Frequency sweep from 10Hz to 55Hz in 90-100 minutes per plane |
| Shock (T4) | 10 fresh fully charged cells<br>Test 4 in sequential test program<br>18 shocks – 150g peak acceleration for 6ms or base don formula and chart is 6th edition of the standard | 5 fresh fully charged cells<br><br>3 shocks – 75g average acceleration, 125-175g peak acceleration |

*Performance - Compliance - Success*

ENERGY ASSURANCE

## IEC 62133 to UL 1642 Thermal Abuse/Heating – Cells

| | IEC 62133 | UL 1642 |
|---|---|---|
| Samples | 5 fresh FC cells prepared at upper temperature<br>5 fresh FC cells prepared at lower temperature | 5 fresh FC cells prepared at upper temperature<br>5 fresh FC cells prepared at lower temperature<br>Same preparation with aged cells may be required |
| Test procedure | 130°C±2°C at a rate of 5°C±2°C per minute<br>Hold for 10 minutes for small cells, 30 minutes for larger cells | 130°C±2°C at a rate of 5°C±2°C per minute<br>Hold for 10 minutes for small cells |
| Pass/fail criteria | No fire and no explosion | No fire, no explosion |
| Other considerations | 30 minute hold for larger cells | Higher oven temp if the cell is specified for temperatures above 100°C |

Potential for an overlapping test – if the following are met:

- Cells are small by IEC 62133 terms – not defined in the standard
- Cells are not intended for above 100°C
- Additional aging test may need to be done for UL 1642

*Performance - Compliance - Success*

ENERGY ASSURANCE

## Battery level testing across standards

| | UN 38.3 | UL 2054 | IEC 62133 |
|---|---|---|---|
| External Short Circuit | X | X | X |
| Abusive Overcharge/Overcharge | X | X | X |
| Moulded case stress/Mold stress | N/A | X | X |
| Free Fall/drop | N/A | X | X |
| Altitude Simulation/Low Pressure | X | N/A | N/A |
| Temperature Cycling | X | N/A | N/A |
| Vibration | X | N/A | N/A |
| Shock | X | N/A | N/A |
| Forced Discharge | N/A | X | N/A |
| Abnormal Charge | N/A | X | N/A |
| Limited Power Source | N/A | X | N/A |
| Component and surface temperature | N/A | X | N/A |
| 250N force | N/A | X | N/A |

*Performance - Compliance - Success*

ENERGY ASSURANCE

## UL 2054 Electrical Tests

UL 2054 Electrical Tests do not overlap with any other test programs

### Faulting

- Removal of safety critical components that are not UL Listed or Recognized:
  - Short Circuit testing
  - Abnormal charging
  - Abusive overcharge
  - Forced Discharge
  - Limited Power Source

### Trip/no-trip testing

- Testing just below the trip point of the safety circuit:
  - Short Circuit testing (faulted)
  - Abnormal charging (faulted)
  - Component and surface temperature test (non-faulted)

*Performance - Compliance - Success*

ENERGY ASSURANCE

## Battery level testing across standards

| | UN 38.3 | UL 2054 | IEC 62133 |
|---|---|---|---|
| External Short Circuit | X | [redacted] | X |
| Abusive Overcharge/Overcharge | X | [redacted] | X |
| Moulded case stress/Mold stress | N/A | X | X |
| Free Fall/drop | N/A | X | X |
| Altitude Simulation/Low Pressure | X | N/A | N/A |
| Temperature Cycling | X | N/A | N/A |
| Vibration | X | N/A | N/A |
| Shock | X | N/A | N/A |
| Forced Discharge | N/A | [redacted] | N/A |
| Abnormal Charge | N/A | [redacted] | N/A |
| Limited Power Source | N/A | [redacted] | N/A |
| Component and surface temperature | N/A | [redacted] | N/A |
| 250N force | N/A | X | N/A |

*Performance - Compliance - Success*

ENERGY ASSURANCE

## External Short Circuit – Batteries

| | UN 38.3 | IEC 62133:2012 |
|---|---|---|
| Samples under test | 4 fresh and 4 cycled batteries<br>Fully charged<br>Previously tested to T1-T4 | 5 fresh charged at low temp –<br>5 fresh charged at high temp –<br>Fully charged |
| Test ambient | 57°C ± 4°C (6hr soak) – NEW in 6th edition | 55°C (±5°C) |
| Circuit Resistance | <0.1Ω | 80±20mΩ |
| Termination Criteria | 1hr or case returns to ambient | 24h or until external temp declines by 20% from the maximum recorded temp rise |
| Pass/fail Criteria | Case temp <170°C, no disassembly, no rupture, no fire during the test and for 6 hours after termination | No fire/No explosion |
| Other considerations | T5 in a sequence of tests | |

*Performance - Compliance - Success*

ENERGY ASSURANCE

# Overcharge – Batteries

| | UN 38.3 | IEC 62133:2012 |
|---|---|---|
| Samples under test | 4 fresh and 4 cycled batteries<br>Fully charged | 5 fresh<br>Fully discharged |
| Charge Current | 2X manufacturer's recommended maximum continuous charge current | $2I_tA$ |
| Charge voltage | 22V (recommended charge voltage <18V) or 1.2X maximum charge voltage | Maximum voltage specified by the manufacturer or 5.0V/cell |
| Termination Criteria | 24hr | Steady state case temperature or return to ambient |
| Pass/fail Criteria | No disassembly and no fire during the test or within 7 days of termination | No fire/No explosion |
| Other considerations | | |

*Performance - Compliance - Success*

ENERGY ASSURANCE

# Mechanical tests – UL 2054 and IEC 62133

| | UL 2054 | IEC 62133 |
|---|---|---|
| Mold Stress/Moulded Case Stress | 3 batteries<br>Fully discharged<br>Test temperature<br>70°C or calculated value based on CST temperature rise data or RTI of plastic +10°C<br>Test duration – 7hr | 3 batteries<br>Fully charged<br>Test temperature<br>70°C±2°C<br>Test duration – 7hr |
| Free Fall/Drop Impact | 3 batteries<br>Fully charged<br>Conditioned at lower temperature and dropped cold if operating range is below 0°C<br>1m to concrete<br>3 drops per sample<br>No cracks that cell or safety circuit are exposed | 3 batteries<br>Fully charged<br>1m to concrete<br>3 drops per sample<br>1 hr wait after drop – no fire and no explosion |

Free fall/drop could overlap if the following are met:

- Operating range is not below 0°C
- Separate criteria for evaluation are used for each standard

*Performance - Compliance - Success*

ENERGY ASSURANCE

## Potential Test overlaps

**Cells**

**Altitude simulation – UN 38.3 and UL 1642**

**And possibly Heating/Thermal Abuse – UL 1642 and IEC 62133**

**Batteries**

**Free fall/drop Impact from UL 2054 and IEC 62133**

*Performance - Compliance - Success*

ENERGY ASSURANCE

## Conclusion

- Not all test standards and tests are created equal.
- However, not all batteries and cells are created equal. A review of each individual construction could lead to additional overlap depending on the construction.

*Performance - Compliance - Success*

ENERGY ASSURANCE

## Thank you!

Energy Assurance LLC

www.Energy-Assurance.com

Cindy Millsaps

cindymillsaps@energy-assurance.com

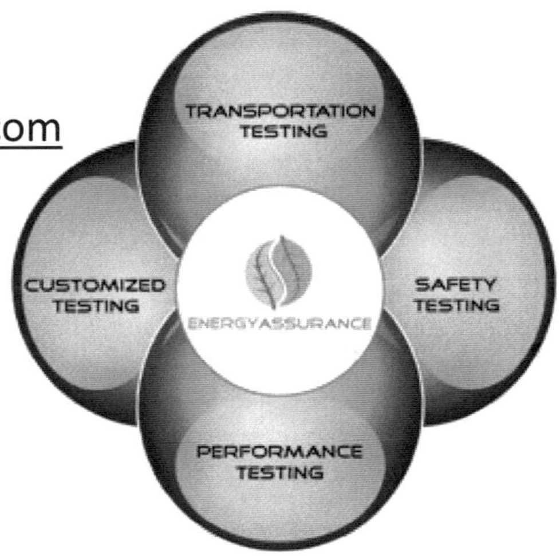


*Performance - Compliance - Success*

Fraunhofer
Heinrich Hertz Institute

Department Fiber Optical Sensor Systems

# Enhanced Battery Safety by applying embedded photonic Sensors

A. Nedjalkov, J. Meyer, M. Angelmahr, W. Schade

Fraunhofer Heinrich Hertz Institute, Am Stollen 19B, 38640 Goslar

www.hhi.fraunhofer.de

## Table of Content

Fiber Optical Sensor Systems





TU Clausthal
Clausthal University of Technology

Fiber Optical
Sensor Systems

# Fraunhofer Heinrich Hertz Institute

© Fraunhofer
Heinrich Hertz Institute

TU Clausthal
Clausthal University of Technology

3

## Departments

Fiber Optical
Sensor Systems

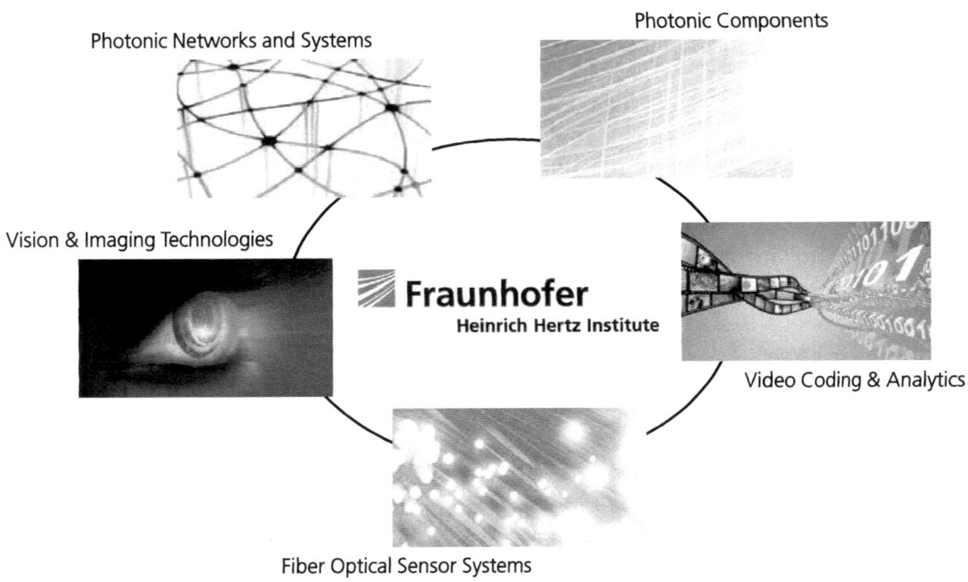


© Fraunhofer
Heinrich Hertz Institute

TU Clausthal
Clausthal University of Technology

4

Fiber Optical
Sensor Systems

Risk of cell destruction caused by
overcharge, aging, external damage or production defects



## Risk of battery cell destruction

Fiber Optical
Sensor Systems

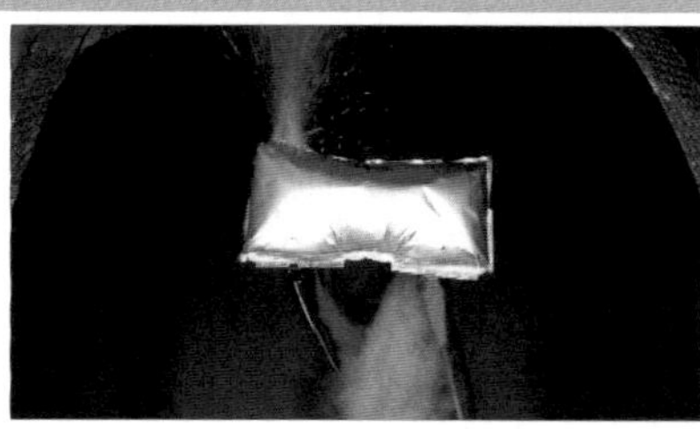

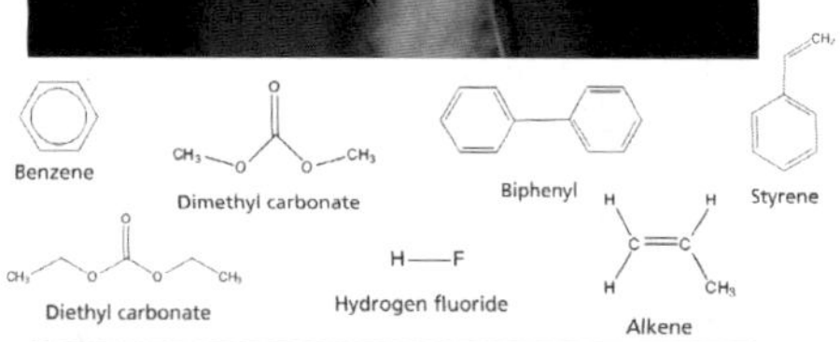

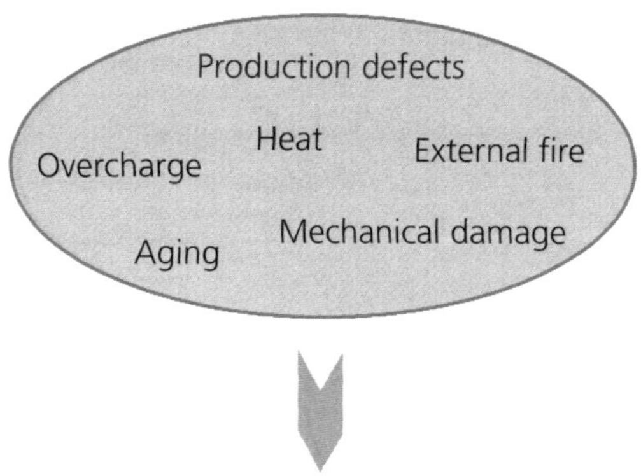


- Carcinogenic and toxic gas emission
- Ignition of flammable battery gases
- Thermal runaway of adjacent cells

## Risk of battery cell destruction

Fiber Optical
Sensor Systems

No switch off of the charge unit because voltage limits were abided

Thermal runaway after failure of the battery charge unit

© Fraunhofer
Heinrich Hertz Institute

TU Clausthal
Clausthal University of Technology

7

Fiber Optical
Sensor Systems

## Battery and Sensor Test Center

© Fraunhofer
Heinrich Hertz Institute

TU Clausthal
Clausthal University of Technology

8

## Battery and Sensor Test Center

Fiber Optical Sensor Systems

Large scale materials processing

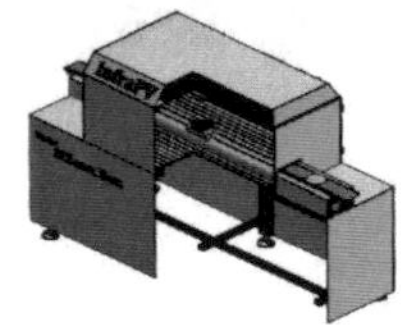

Batteries in the extreme

Smart sensors for industry 4.0

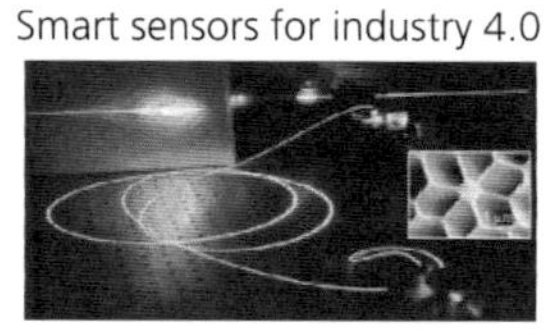



## Charging disruption with photonic monitoring

Fiber Optical Sensor Systems

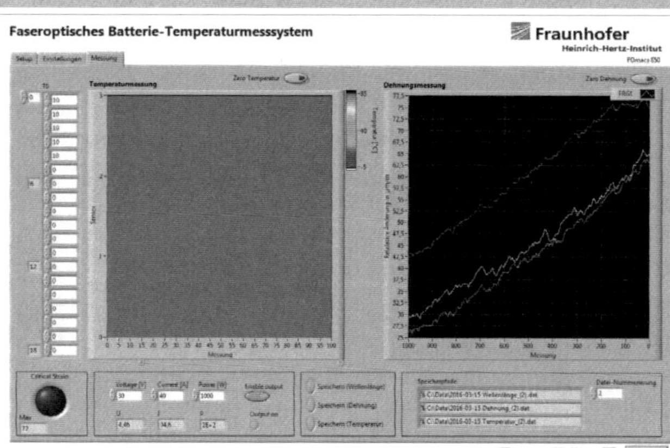


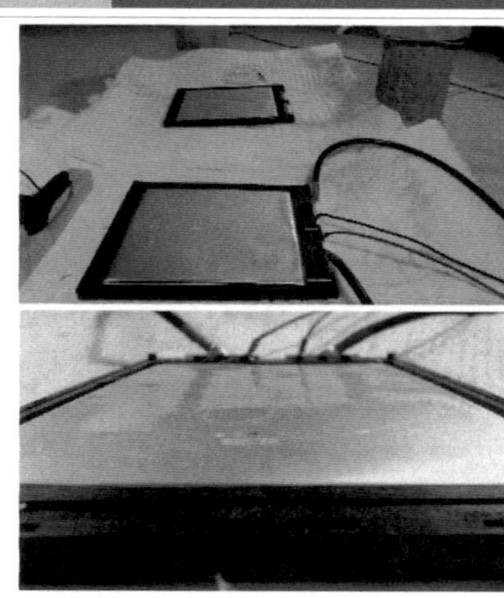

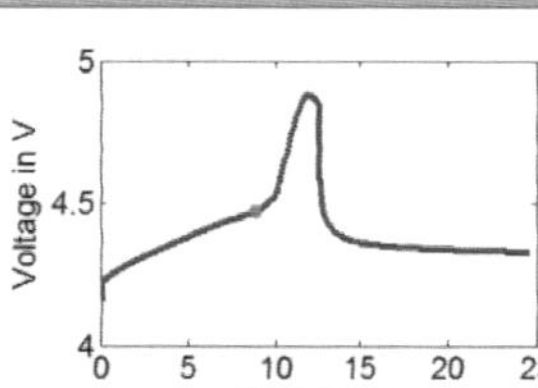


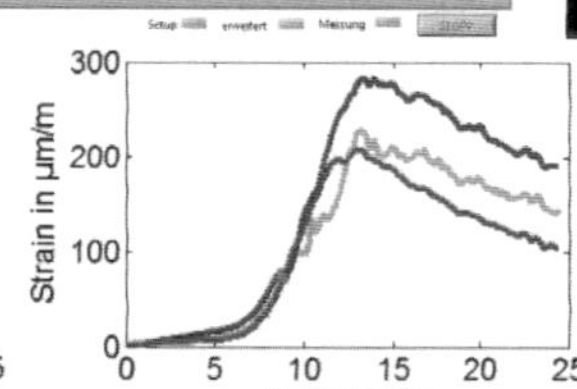


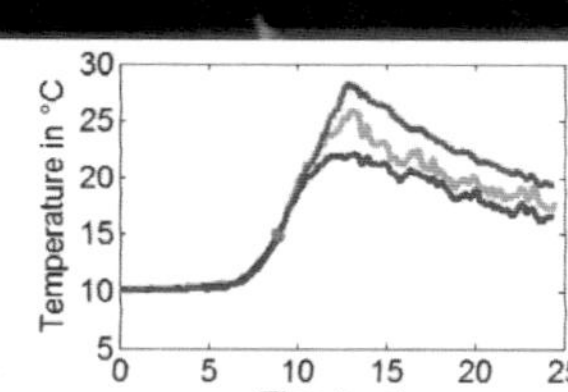

# Disabled photonic monitoring

Fiber Optical
Sensor Systems

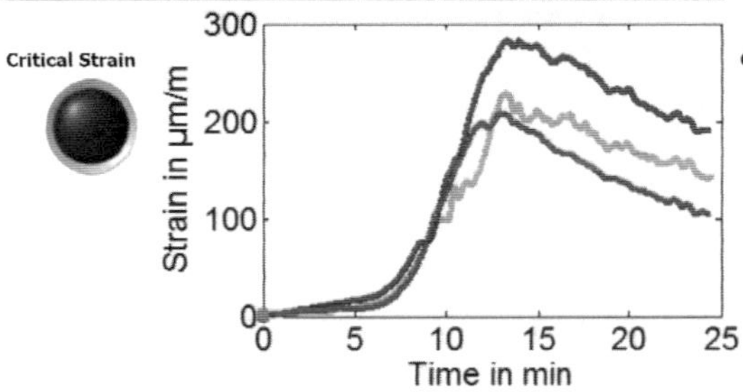


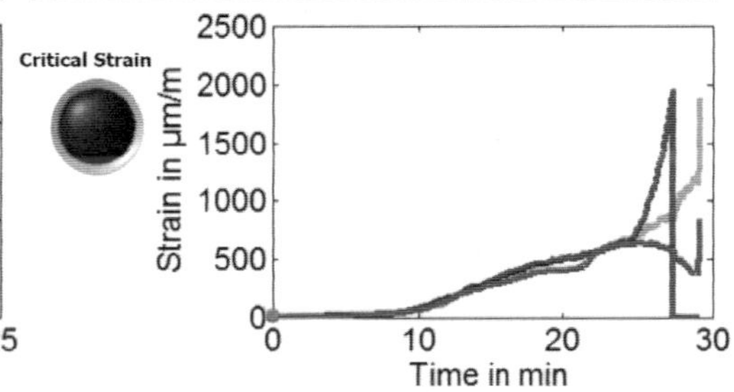




Fiber Optical
Sensor Systems

## Battery safety enhancement with fiber optical sensors

## Operating principle fiber-Bragg-granting

Fiber Optical Sensor Systems

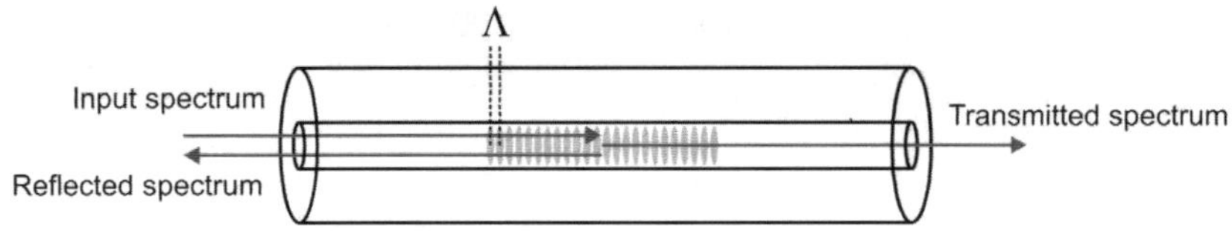


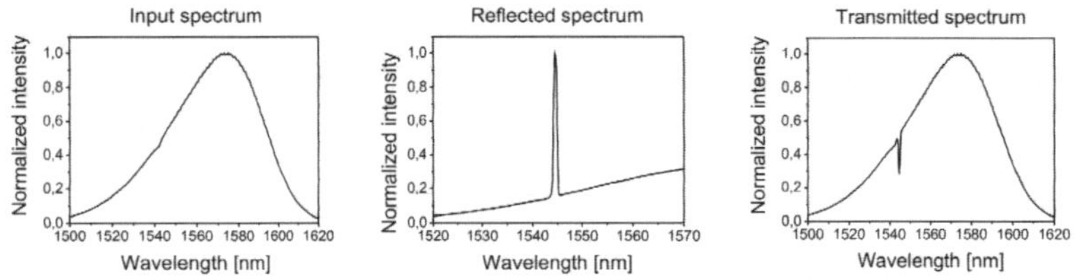


**Bragg condition**

$$\lambda_B = 2n_{eff}\Lambda$$

$\Lambda$: grating period

$n_{eff}$: effective refractive index



## Ultimate goal: Lab-in-a-Fiber (LIF)

Fiber Optical Sensor Systems

Optical fiber, which combines multiple sensor tasks

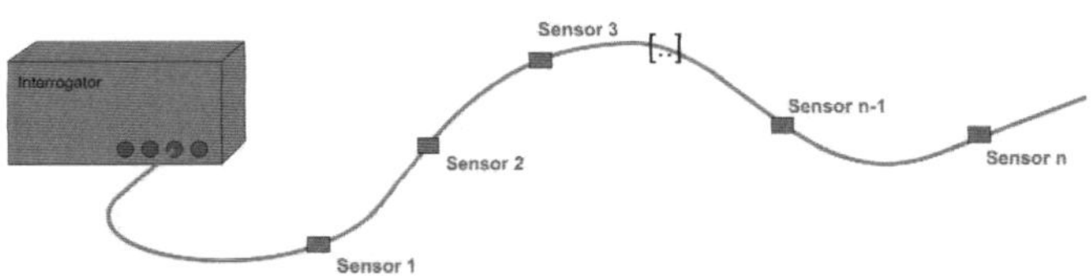


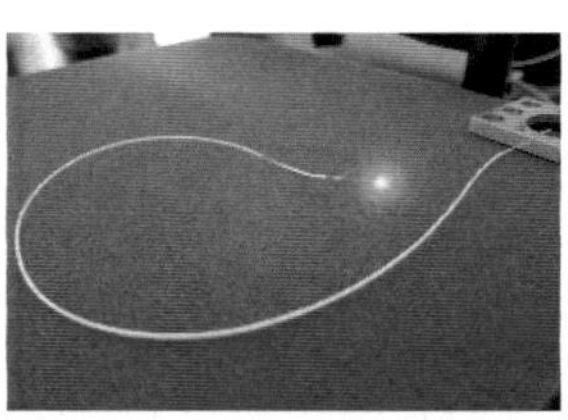

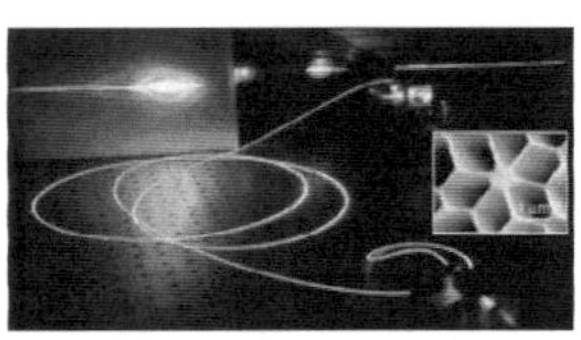

Physical properties to be measured:

- Temperature
- Strain and 3D profile
- Pressure / Acoustics
- Chemical composition
- Voltage and current

Evanescent guided light

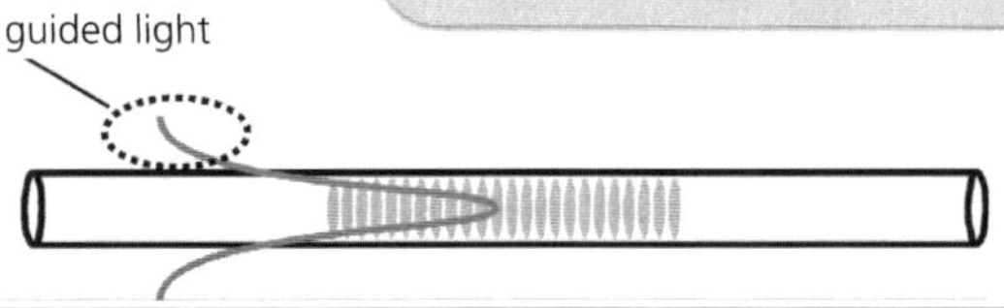

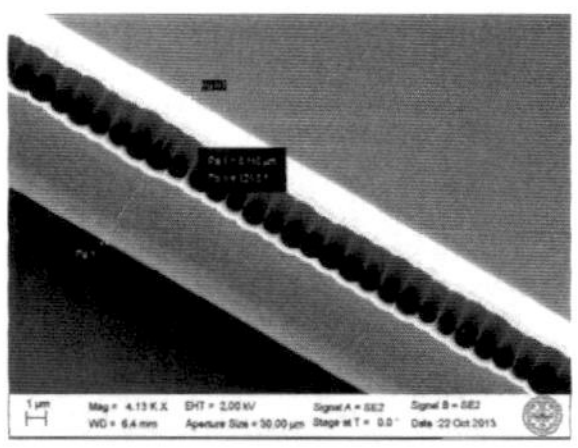

## Ultimate goal: Lab-in-a-Fiber (LIF)

Fiber Optical
Sensor Systems

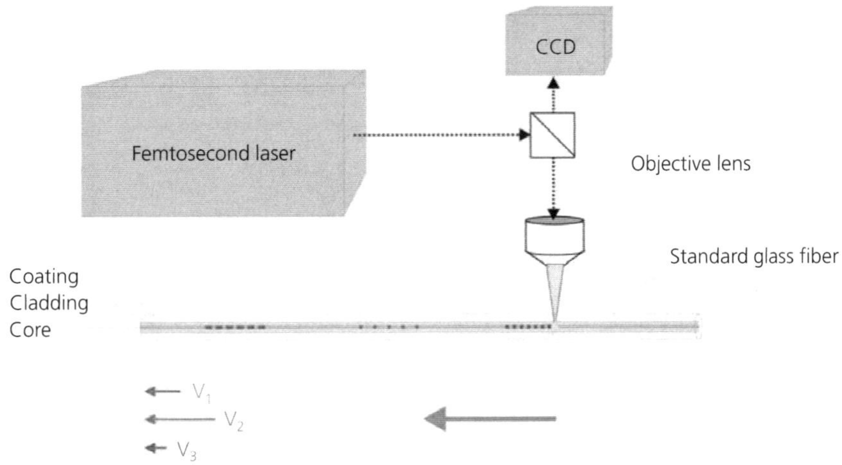


Point-by-point Femtosecond laser inscription of fiber Bragg gratings



## Application of FBG in battery storage systems

Fiber Optical
Sensor Systems

Distributed temperature and strain sensing

E. g.

**9 temperature** and
**1 strain sensor**

for each battery cell are combined in a single fiber optical string. (I.)

Each sensor is characterized by the specific wavelength of its FBG. (II.)

Within a battery stack a single cell monitoring is realized. (III.)

I.

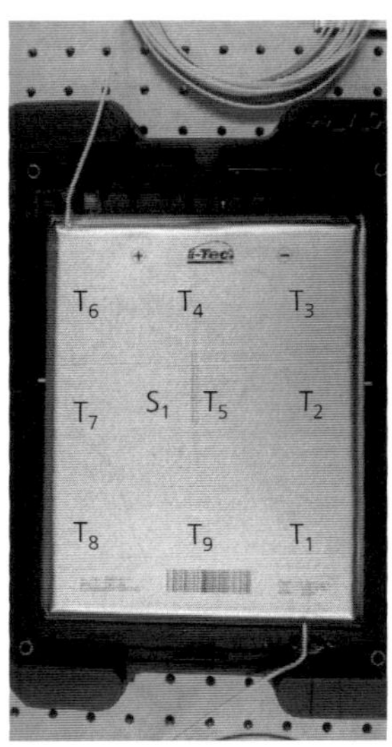


II.

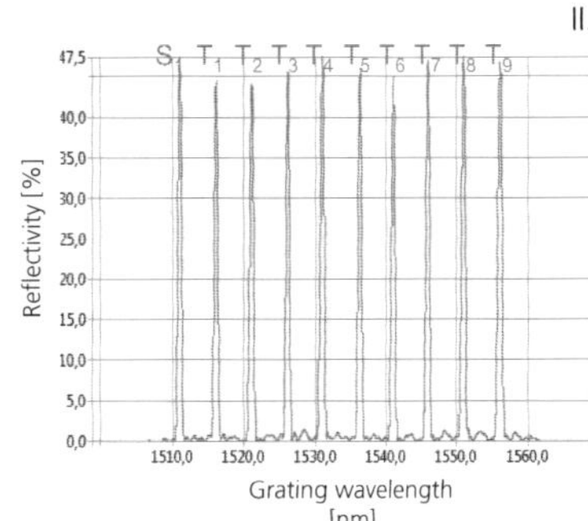


III.

## Battery safety enhancement – Temperature field monitoring during fast charging

Fiber Optical Sensor Systems

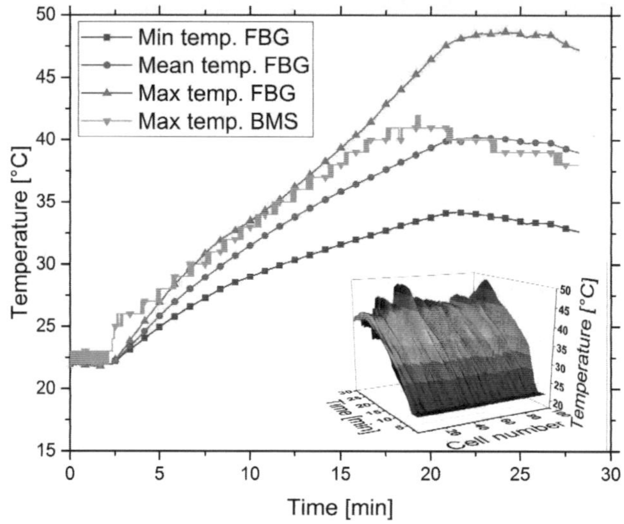


Dissimilarity between fiber optically and conventionally measured temperatures exemplified by a fast charged battery system of an electric vehicle



## Battery safety enhancement – strain and temperature during overcharge and nail test

Fiber Optical Sensor Systems

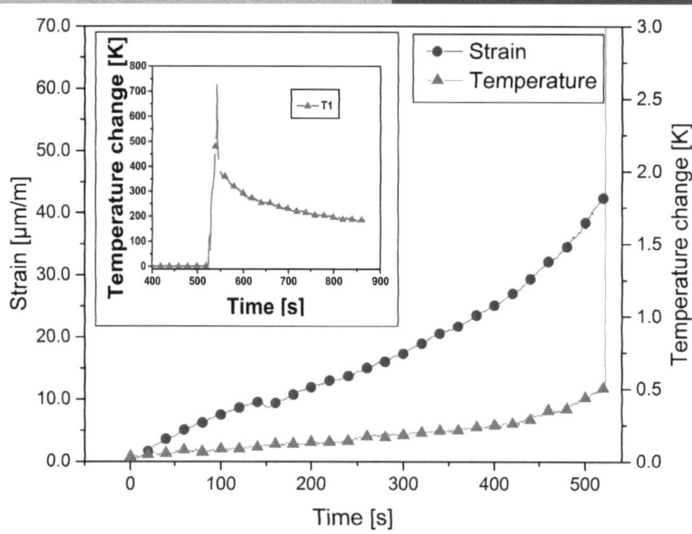


I. Overcharging of a previously fully charged battery cell (40 Ah) with a current of 20 A
II. Subsequently performed nail test, as required by the customer

## Battery safety enhancement – irreversible long-term strain behavior

Fiber Optical Sensor Systems

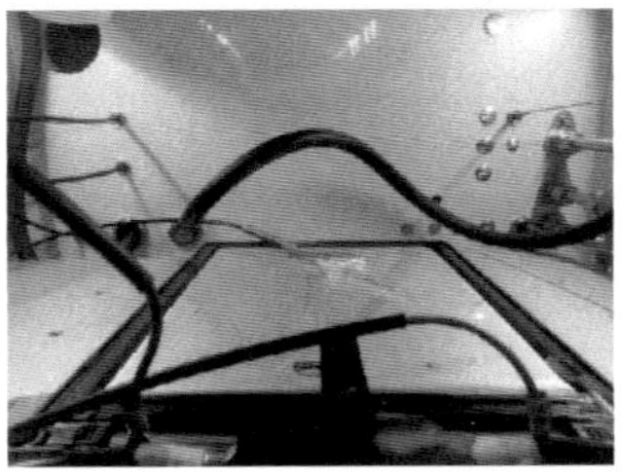

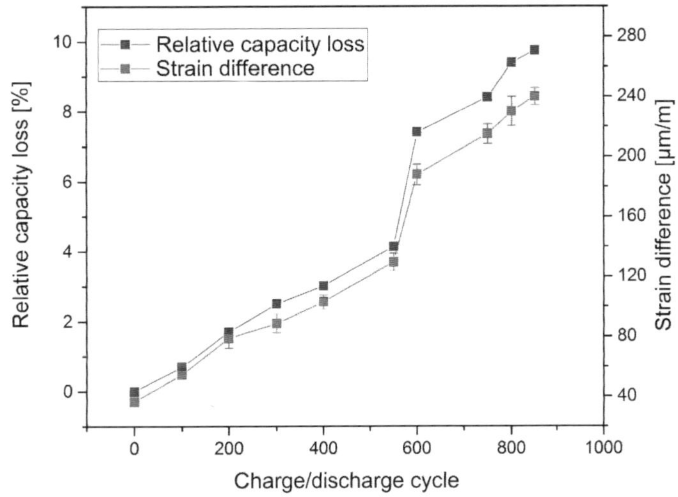


Accelerated aging due to increased ambient temperature (40°C) at a cyclization rate of 1C



## Battery safety enhancement —cycling behavior for a stationary application

Fiber Optical Sensor Systems

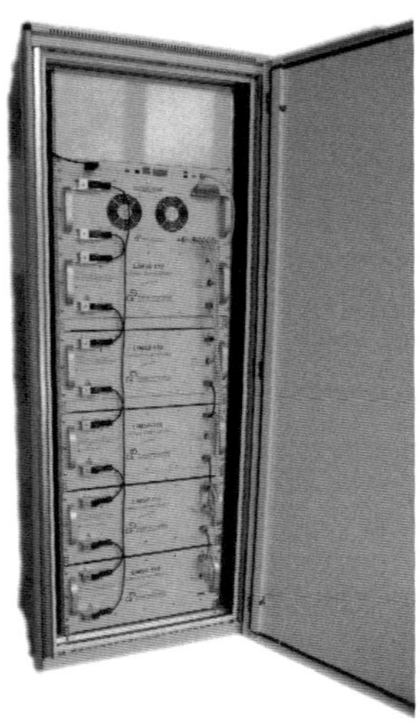

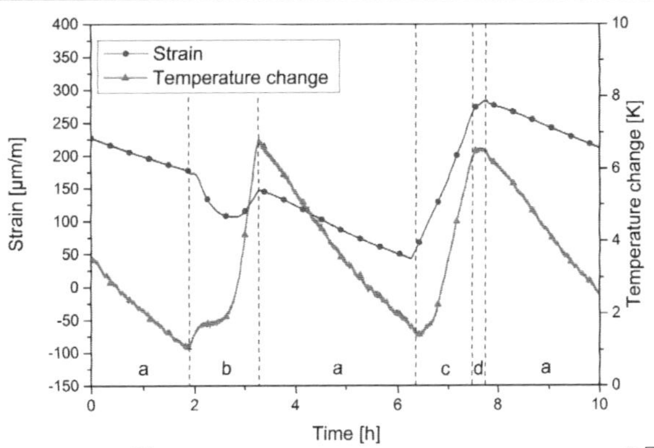


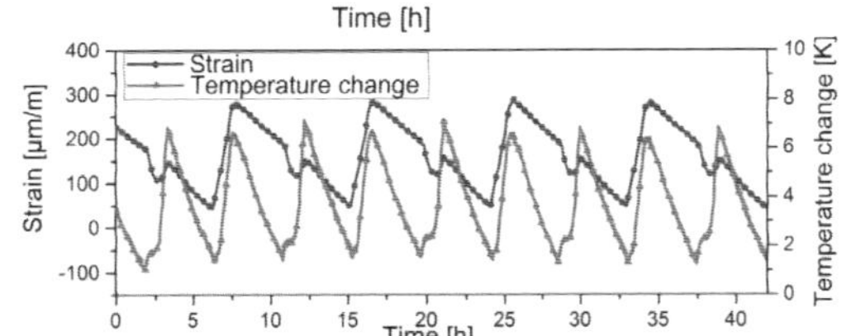


Operation of a battery storage cabinet under standard conditions with the sections interphase (a), discharge (b), CC-charge (c) and CV-charge (d)

## Battery safety enhancement – application benefits of fiber optical sensors

Fiber Optical Sensor Systems

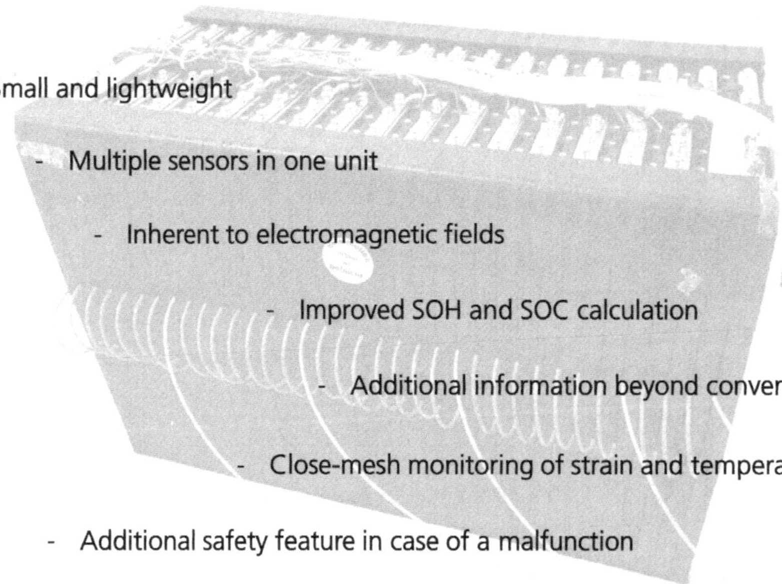

- Small and lightweight
- Multiple sensors in one unit
- Inherent to electromagnetic fields
- Improved SOH and SOC calculation
- Additional information beyond conventional BMS
- Close-mesh monitoring of strain and temperature
- Additional safety feature in case of a malfunction



## Certified fire protection cover for li-ion batteries in stationary applications

Fiber Optical Sensor Systems

Prospective strict requirements for stationary battery applications by progressive standardization and certification

Hazards to be prevented !

Thermal runaway of single cells
Cascade propagation of thermal runaway
Poisoning due to the spewing of hazardous battery gas substances
External fire by accompanying appliance
Internal fire by electronic components

Multilevel security system for risk minimization in the use of lithium ion batteries

## Certified fire protection cover for li-ion batteries in stationary applications

**Fiber Optical Sensor Systems**

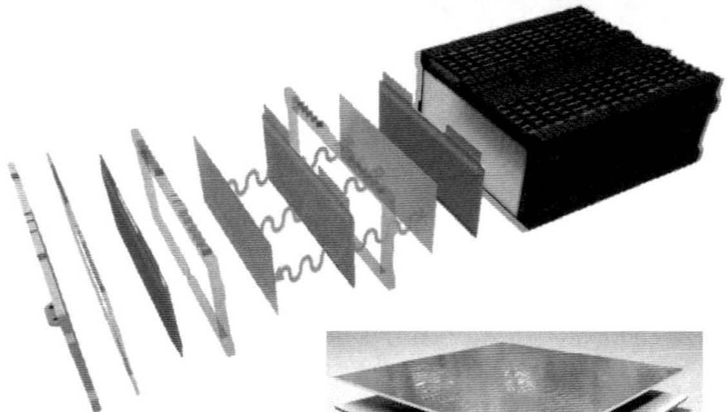

I. Individual cell monitoring by fiber optic sensor units

II. Energy absorption by cell interlayers in case of a cell damage

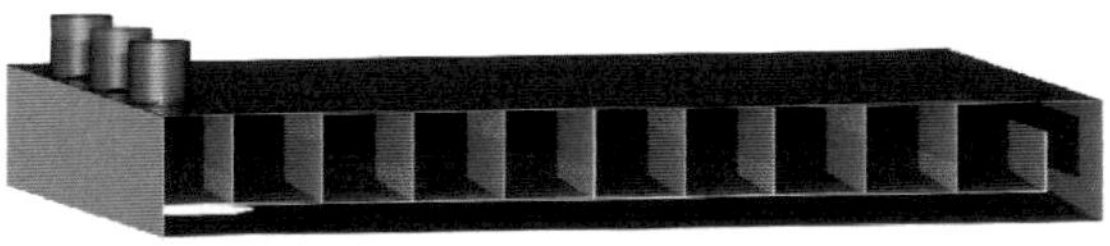

III. Meandered gas filtering system filled with neutralization chemicals

© Fraunhofer Heinrich Hertz Institute · STÖBICH technology · KTeX · Power Innovation Stromversorgungstechnik GmbH · TU Clausthal 23

## Certified fire protection cover for li-ion batteries in stationary applications

**Fiber Optical Sensor Systems**

IV. Fire-protected battery energy storage cabinet

V. Fire protection system

VI. Gas-tight battery storage case

© Fraunhofer Heinrich Hertz Institute · STÖBICH technology · KTeX · Power Innovation Stromversorgungstechnik GmbH · TU Clausthal 24

Fiber Optical
Sensor Systems

Future generation of photonic monitoring

© Fraunhofer
Heinrich Hertz Institute

TU Clausthal
Clausthal University of Technology

25

## Optical waveguide integrated sensor foil

Fiber Optical
Sensor Systems

Optical Concept for strain field and temperature field measurements

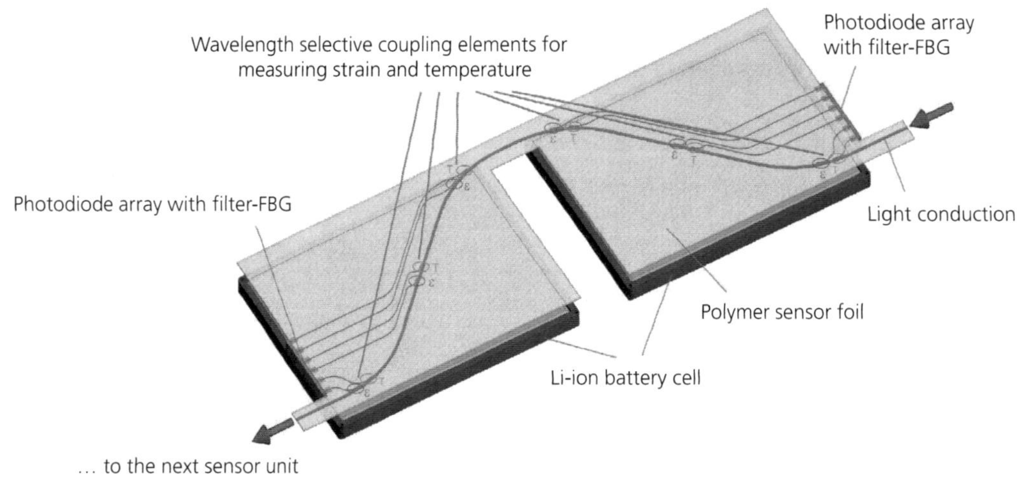


© Fraunhofer
Heinrich Hertz Institute

TU Clausthal
Clausthal University of Technology

26

# Optical waveguide integrated sensor foil

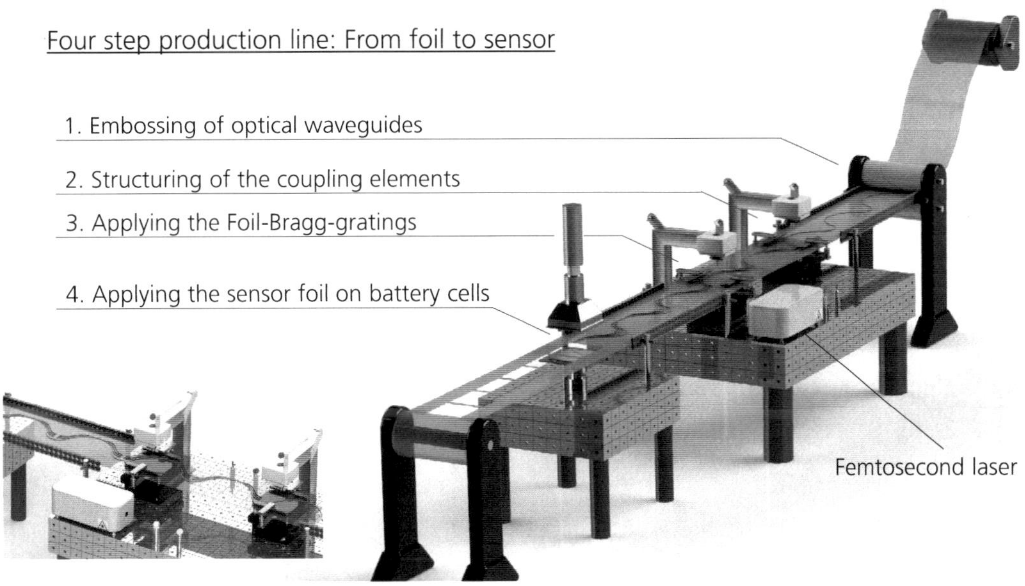






## Department Fiber Optical Sensor Systems

# Department Fiber Optical Sensor Systems

Fiber Optical
Sensor Systems

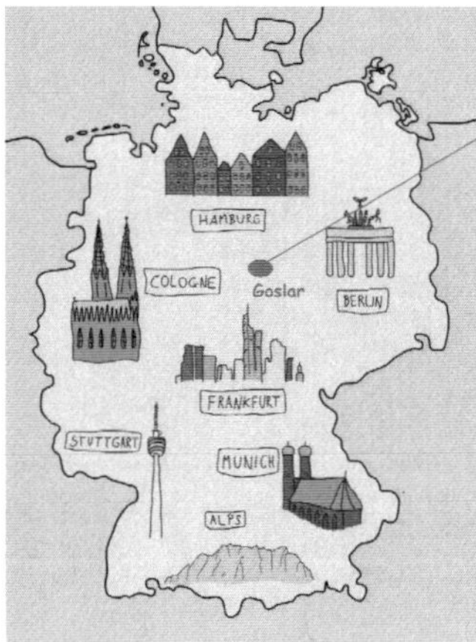

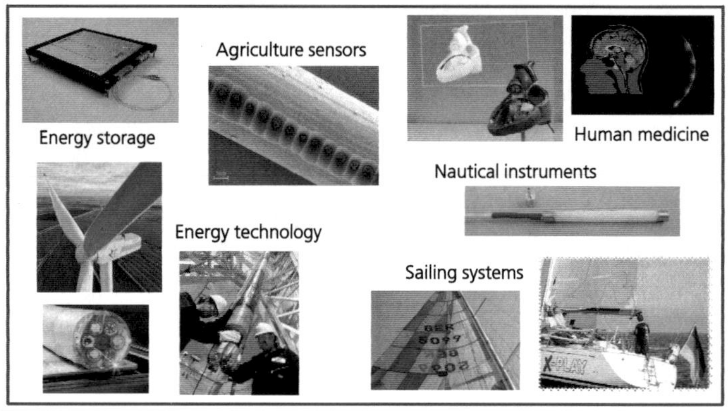




# Contact

Fiber Optical
Sensor Systems

Head of Department
Fiber Optical Sensor Systems
Prof. Dr. Wolfgang Schade
wolfgang.schade@hhi.fraunhofer.de
+49 5321 3816 8410

Head of Group
Fiber Optical Microsensors
Dr. Martin Angelmahr
martin.angelmahr@hhi.fraunhofer.de
+49 5321 3816 8404

Scientific Employee Associate
Fiber Optical Microsensors
Antonio Nedjalkov
antonio.nedjalkov@hhi.fraunhofer.de
+49 5321 3816 8408

# Safe Lithium-ion Battery Designs for Use, Transportation and Second Use

Judy Jeevarajan, Ph.D.
Underwriters Laboratories Inc.

March 2016
Battery Safety Symposium
International Battery Seminar 2016
Ft. Lauderdale, FL



## Introduction

- Lithium ion batteries are being used in ground, aviation, space, sea, etc. applications in various sizes
- Introduced in recent years into the utility/stationary energy storage industry
- Energy Storage Industry deployed 40.7 MW of capacity in Q2 2015 (9 times compared to Q2 2014)
  - Other cumulative installations totalled 5.6 MW
  - Total ~ 50 MW in 2015
  - Goal: Build or Purchase 1.3 GW of storage by 2020 in California
- Automotive batteries are being repurposed for utility/stationary storage applications
- With increasing energy density at the cell level and with the need for high voltage, high capacity, high power battery systems, questions arise on the safety of the battery systems
  - Has a detailed FMEA/FTA been created for the energy storage system?
  - Have all the system level hazards been addressed?
  - Has extensive testing been carried out at all levels to confirm that safety controls designed at each level work as expected – cell, module, battery, etc.
  - Does the safety change with other factors?
    - Size of battery?
    - Environments (temperature, pressure, dust, humidity, vibrations, etc.)?
    - Cycle and calendar life (including storage)?
    - Special transportation needs and safety during transportation?
    - Other unique requirements?

2

## Growth in Size of Batteries used for Various Applications

**Challenges:**

- **Cell to Module to Battery Level Scale-Up**

Cell level controls do not necessarily translate to module or battery level controls

- **Testing in the Relevant Environment**

All safety controls need to be verified by testing at the appropriate level and in the relevant environment

Example: Hazards such as overcharge and external short have opposite outcomes in pressurized versus non-pressurized environments due to the difference in heat dissipation

3

## Battery Manufacturing Process Challenges

- Tens of thousands to billions of cells manufactured for different types of applications from portable equipment to large ESS.
- Challenge is to **screen and match** every individual cell.
  - Typical COTS and some custom battery manufacturing process does not include cell screening and matching.
  - Cells are assembled into batteries in the 'as received" condition at lower SOC (typically 40%)
- Are assembled batteries **tested under relevant stringent** conditions before sent out into the field?
- Battery **certification tests** that are **comprehensive enough** with **stringent pass/fail criteria**
- Certification tests that verify the **operation of the safety controls** designed into the battery

4

## Challenges with Large Battery Designs

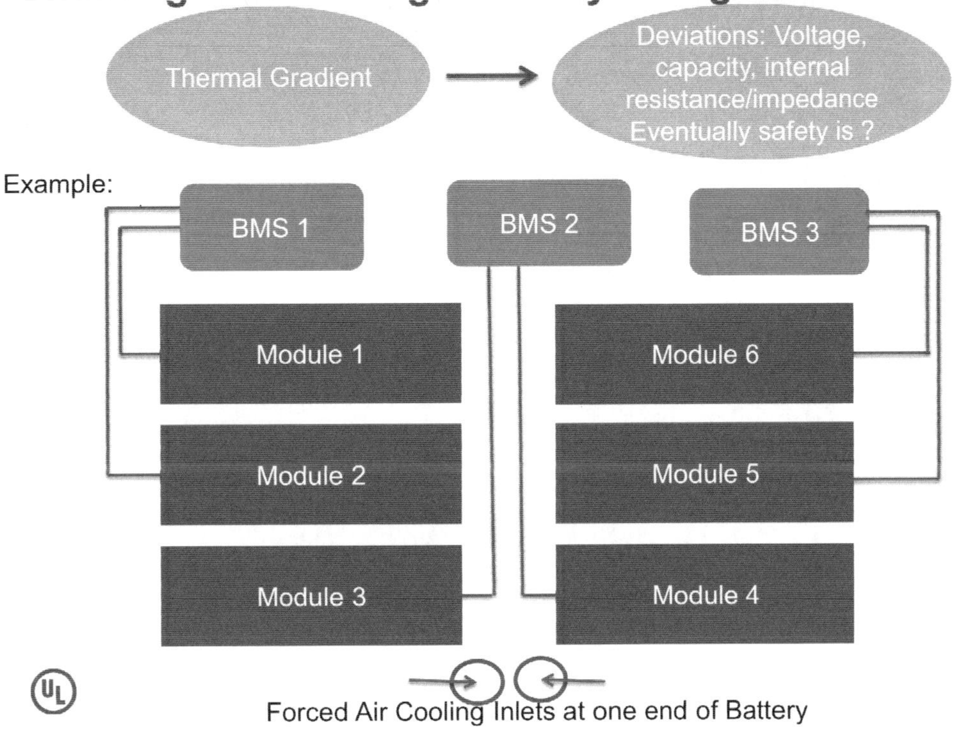


5

## Other Challenges

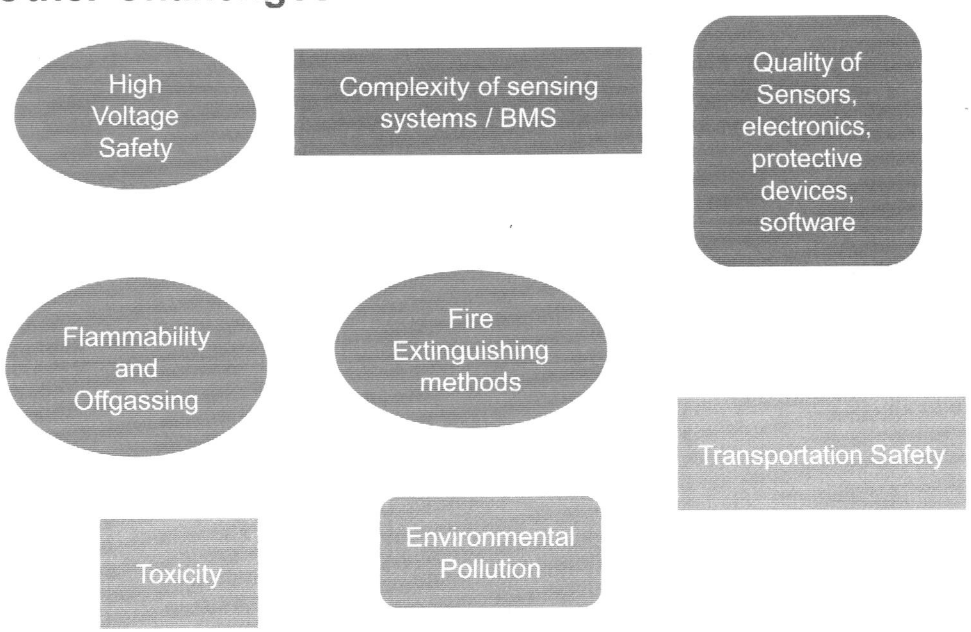


6

## Lithium-ion Batteries: Hazards

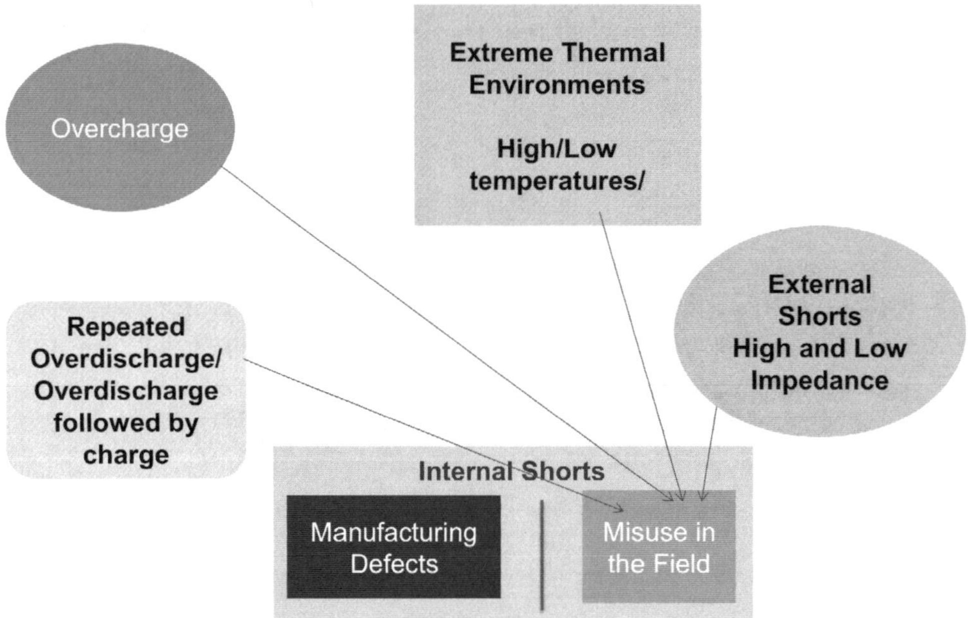


7

## Overcharge and Overdischarge Hazard Causes

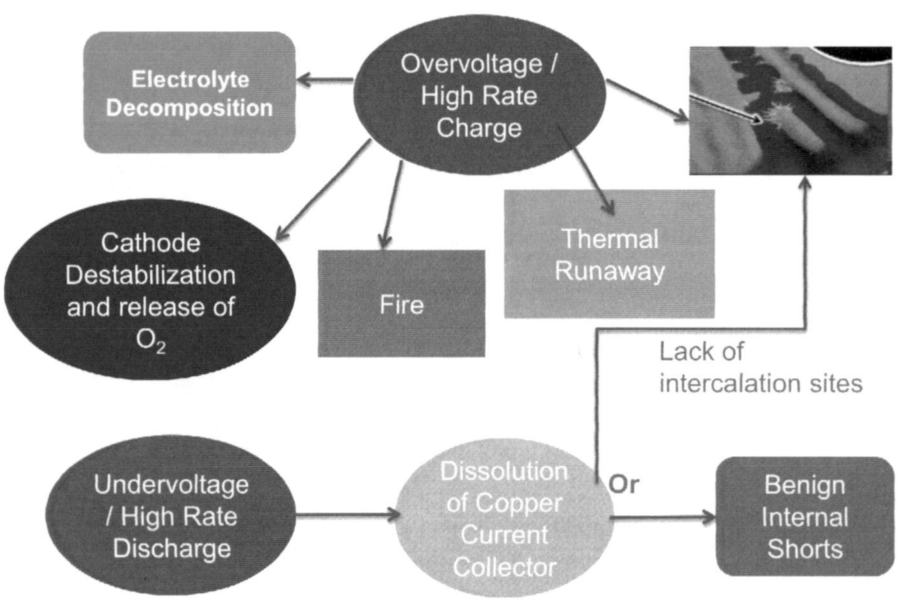


8

# Cell Protective Devices and Limitations

**Lithium-ion cells, whether cylindrical, prismatic, etc. irrespective of size, have different forms of internal protective devices**

- **PTC**
- **CID**
- **Tab/lead meltdown**
  - **(fusible link type)**
- **Bimetallic disconnects**

**-etc.**

**18650 Cell Cross-Section**

**External protective devices used in lithium-ion battery designs are**

- **Diodes**
- **PTC/polyswitch/contactors**
- **Thermal fuses (hard blow or resettable)**
- **Circuit boards with specialized wire traces**

**etc.**

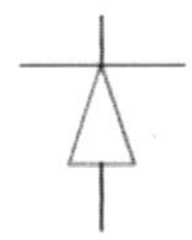

**Diode**

**CID**

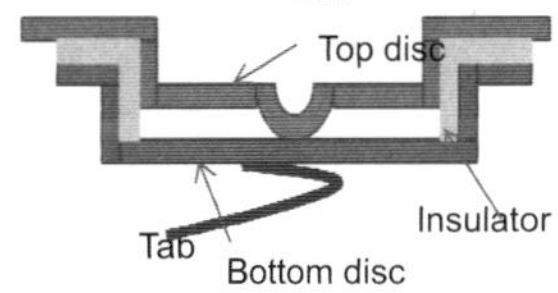


Publications:

1. J. A. Jeevarajan, *et. al.*, "Limitations of Internal Safety Devices in Lithium-ion Cylindrical Cells in Multi-Cell Configurations", Proceedings of the 43rd Power Sources Conference, July 2008.
2. J.A. Jeevarajan, *et. Al.*, "Safety of COTS Lithium-ion Cylindrical Cells in Multi-Cell Configurations in a Vacuum Environment", Proceedings of the 2008 NASA Battery Workshop, November 2008.

9

# Overcharge Test on Single 18650 Cells and High Voltage (14S) String

**Single Cell Test- no thermal runaway**

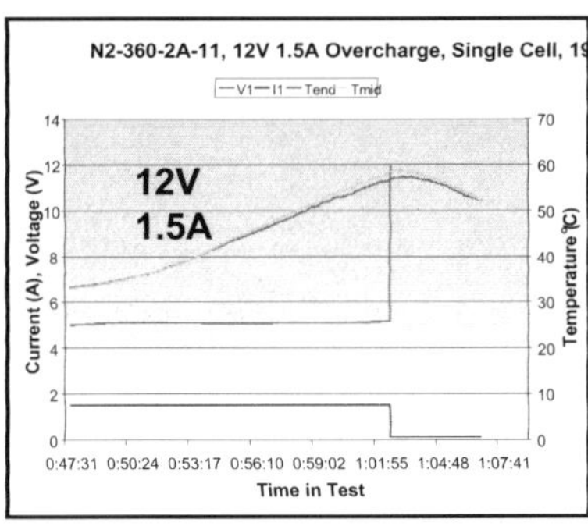


| | CID Open Time (approximate) | Sample Temp. at CID Open (°C) |
|---|---|---|
| N2-360-2A-11 | 1:02:38 | 59 |
| N2-360-2A-12 | 1:02:28 | 73 |

CID Activation - occurs reliably

**14S String Test – Thermal Runaway**

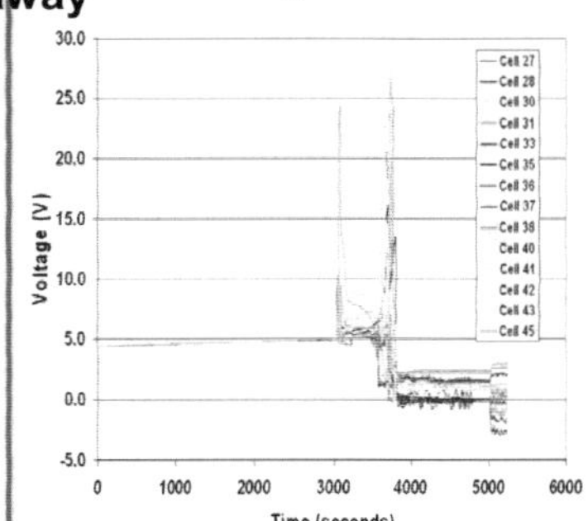

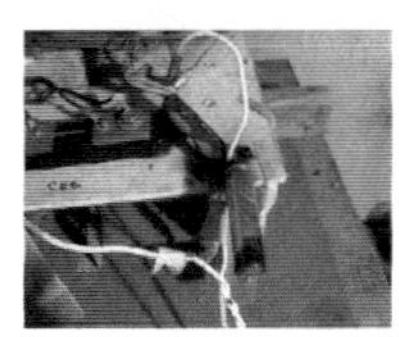

Missing : 27, 28, 30 and 31
Cells 37, 38 and 45 showed no visible Signs of venting

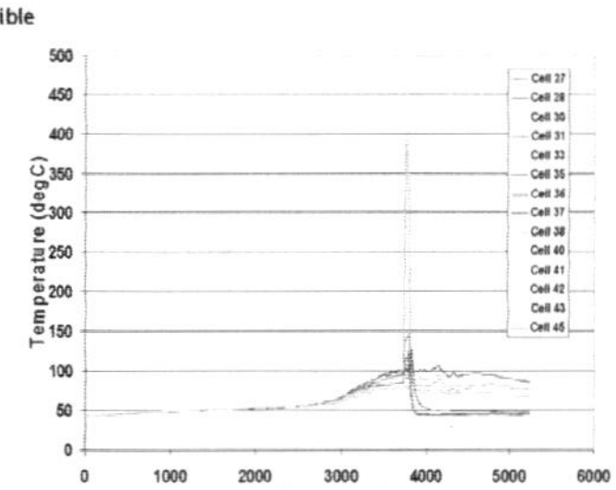


10

# External Short Circuit Hazard

Electrical shock to the cell or battery from external sources.
Usually short circuit of very low resistance is observed.
Very high temperatures are observed.
Venting and fire are also observed.

**18650 Hard Carbon Cell – with PTC**

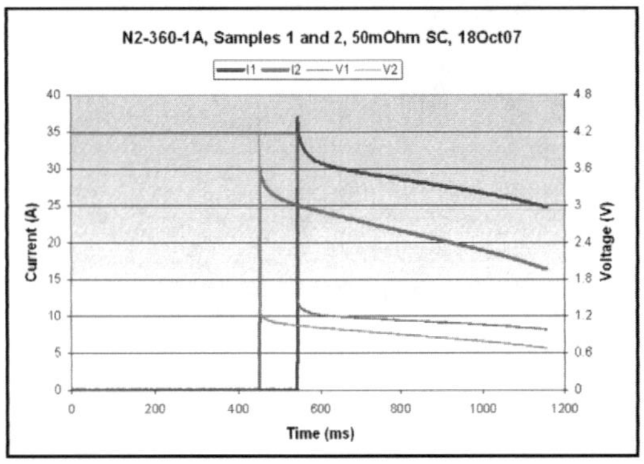


PTC Activation - occurs reliably

**18650 Spinel Cell – without PTC**

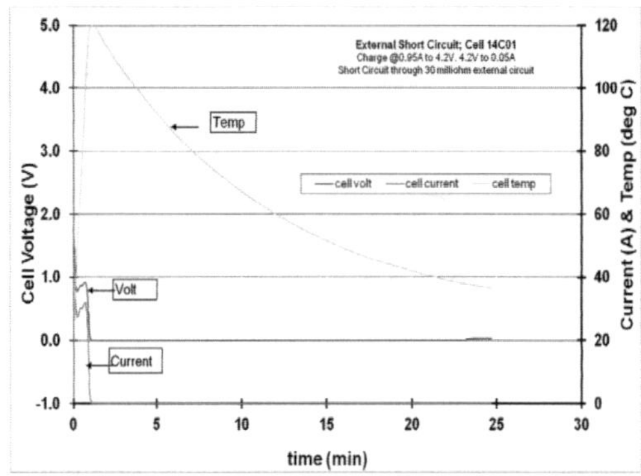


No venting, fire or thermal runaway

11

# Multi-Cell Module Short Circuit Test on Li-ion Cells

**18650 Hard carbon cells with PTC (14S config.)**

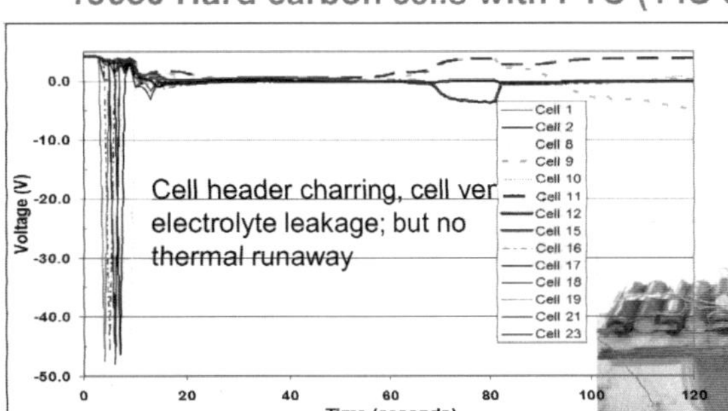


Cell header charring, cell vent, electrolyte leakage; but no thermal runaway

Only venting observed, no fire
Cell Voltages in bold – No visible signs of Venting or damage, but no voltage;
All others vented with Obvious leakage or discoloration

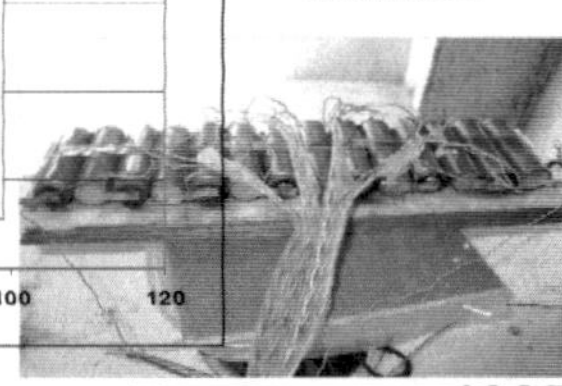

Rapid cell venting within the first 10 seconds

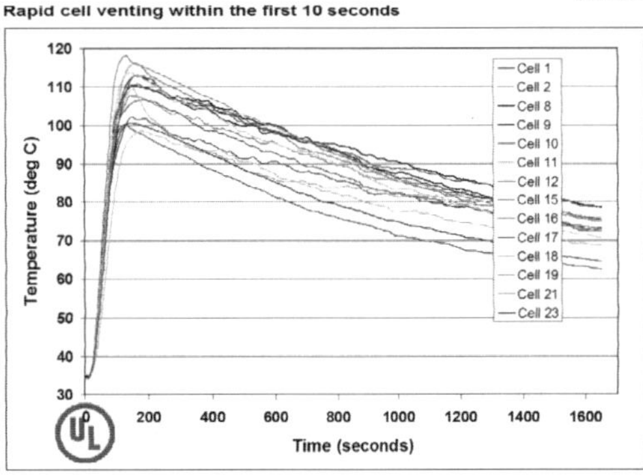


**18650 Spinel Cells - without PTC (10P config.)**

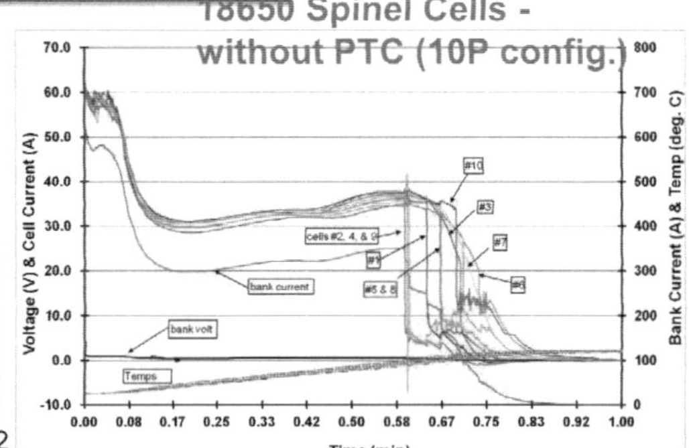


12

## Internal Short Circuit Hazard

Short circuit that occurs inside a cell is called internal short circuit

Usually occurs due to defects inside the cell causing breakage of separator and consequently short circuit

Or it can be caused if the cell is used outside the manufacturer's specification.

High temperatures, venting and fire are observed.

Internal Short

- **Manufacturing Defect**
- **Field Failures**

**Manufacturing Defect**

Major defects screened by manufacturers

<u>**Subtle defects** need to be identified and screened out during acceptance testing (for space applications these are called flight acceptance tests).</u>

**Field Failures**

Avoided by **use within manufacturer's specification** (I, V, T); **stringent cell and battery selection and screening criteria; stringent monitoring and control (I, V, T); cell balancing, health checks** (with issue- recognizable tests); **good thermal design**

13

## Nature of Defects Commonly Observed with Li-ion Cells

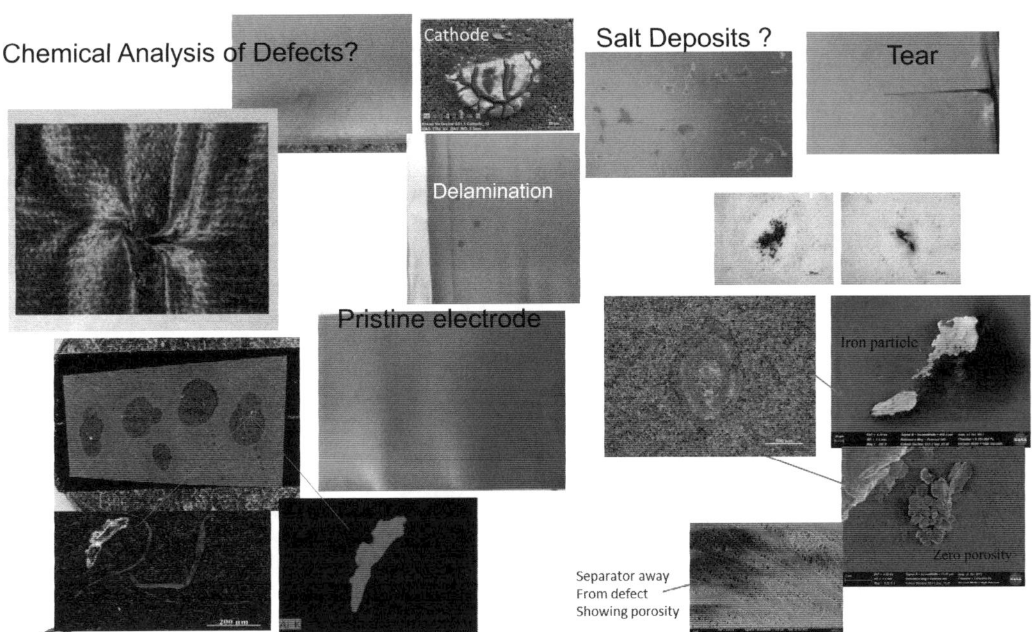


14

## High and Low Temperature Hazards

High Temperatures:
Electrolyte decomposition and gas production
Cathode and anode destabilization
Can lead to venting and fire.

Low Temperatures:
Electrolyte viscosity increases
Increases resistance for the flow of ions
Can result in lithium metal dendrite formation

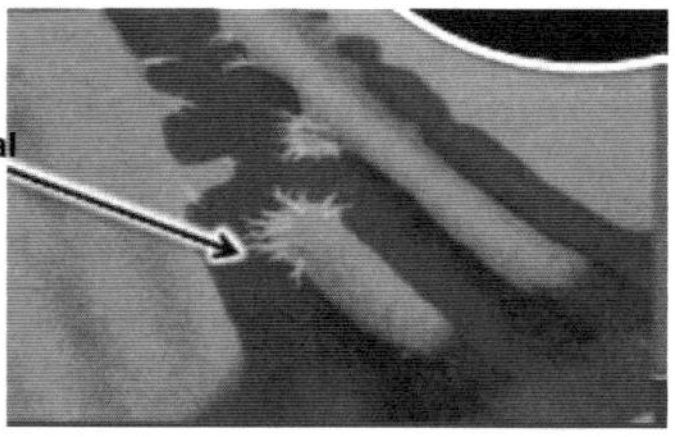


15

## Pouch Cell Studies –Overcharge – single cell versus module (9S & 8P)

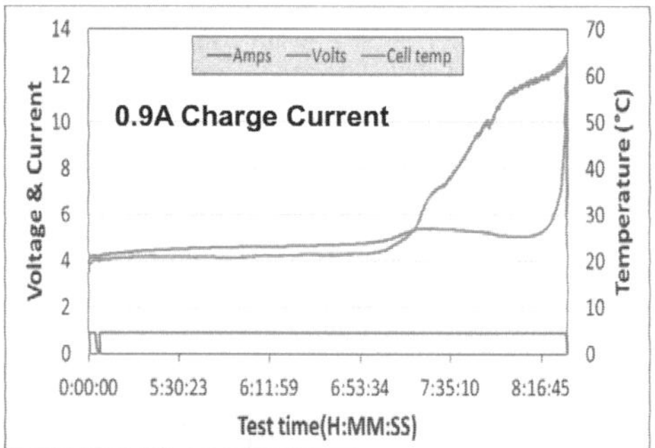


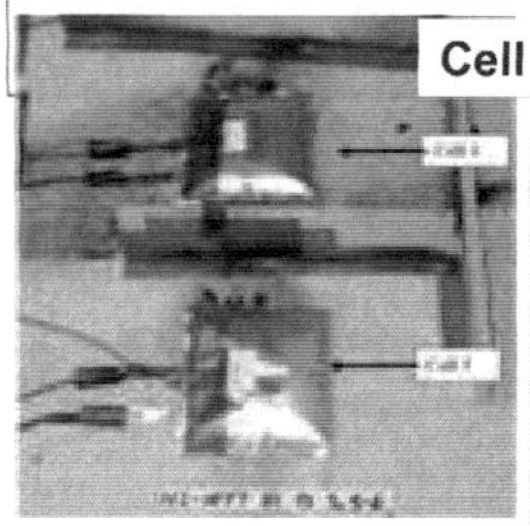


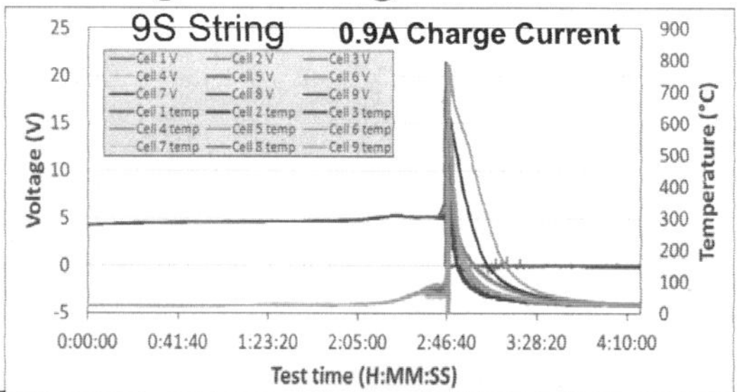


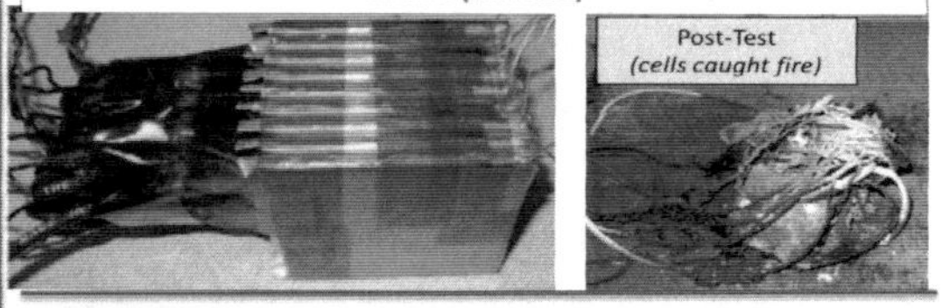


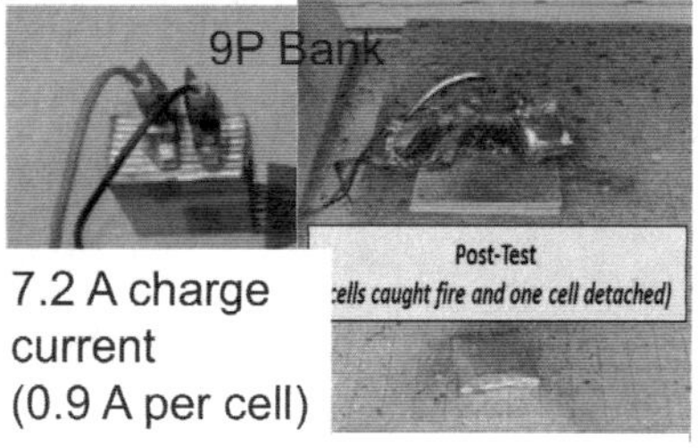


7.2 A charge current
(0.9 A per cell)

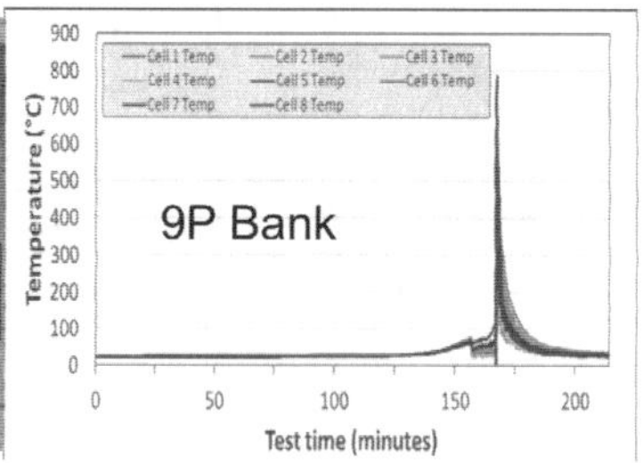

## Pouch Cell Studies –Ambient; Vacuum and Low Pressure Environments

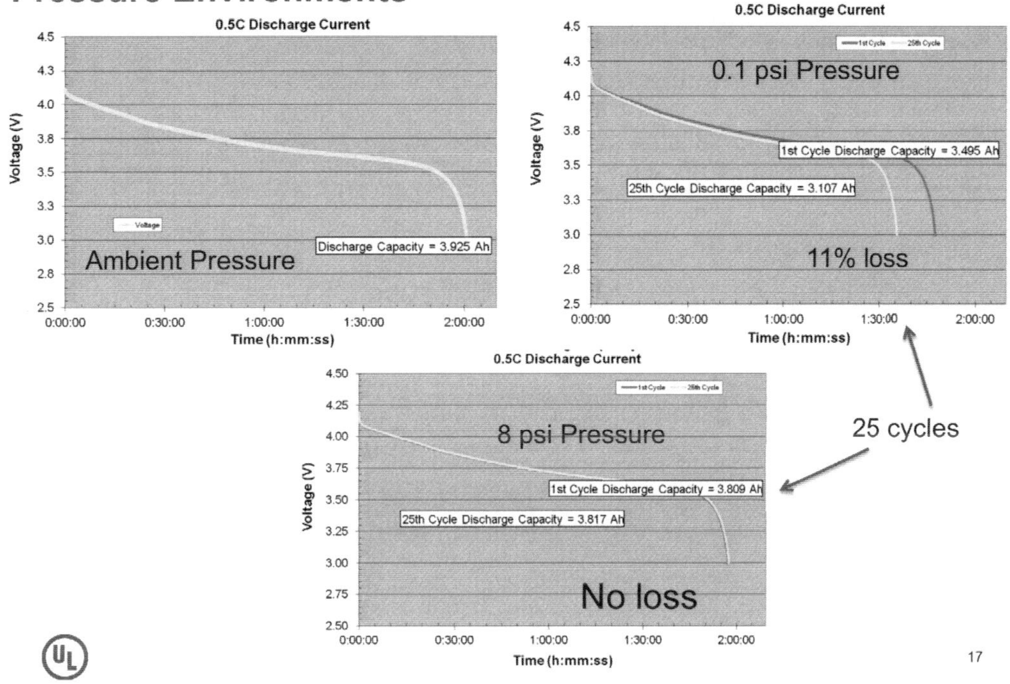


17

## Factors Affecting Aging and State of Health

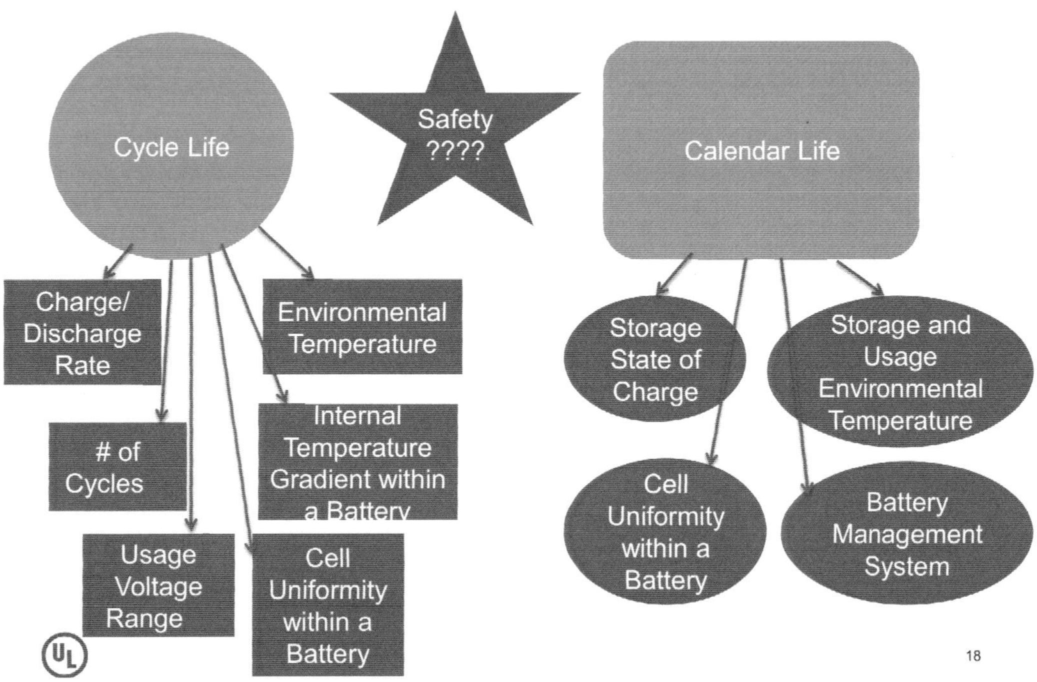


18

## Aging Effects on Cell

- **Lithiation and de-lithiation causes**
  - Anode electrode morphology changes and volume changes – surface can form cracks leading to electrical isolation; delamination from current collector; changes in intercalation kinetics; loss of active lithium inside anode, etc.
- **Decomposition**
  - Binder and electrolyte; SEI decomposition; HF production, li-ion side reaction with electrolyte; etc.
- **Corrosion**
  - Current collector, cell can materials, pouch cell swelling and shorting due to corrosion of pouch material, etc.
- **Cathode changes**
  - Structural disorder, metal dissolution, disproportionation, etc.

19

## Cycle Life Performance at Different Rates

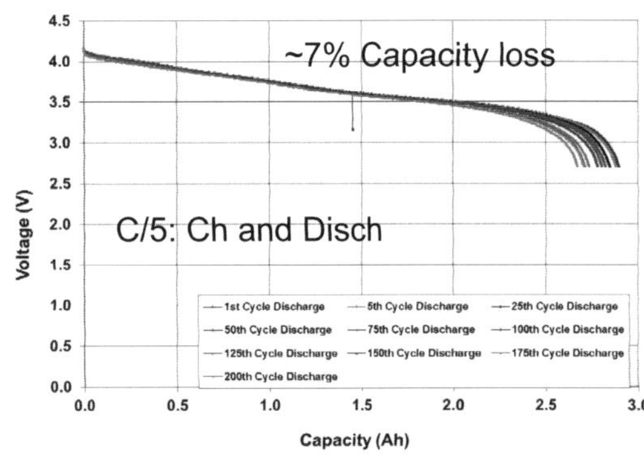


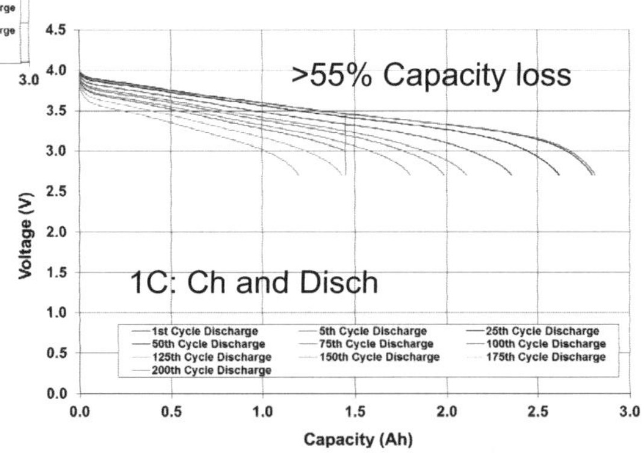


J. Jeevarajan, NASA Biennial Research Publication; 2011

20

## Capacity Loss and Internal Resistance Growth for Cells Used in Orbiter Upgrade Study

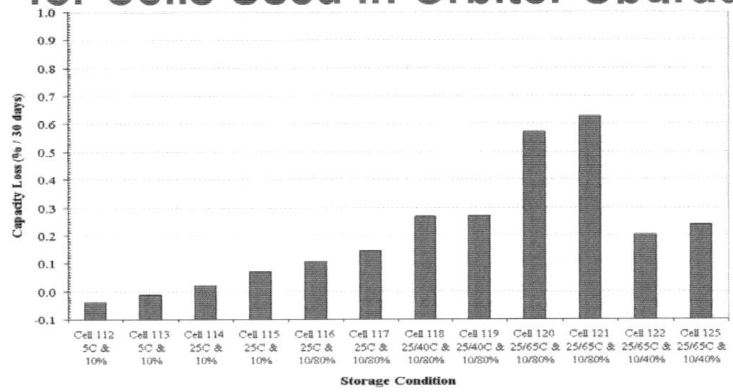


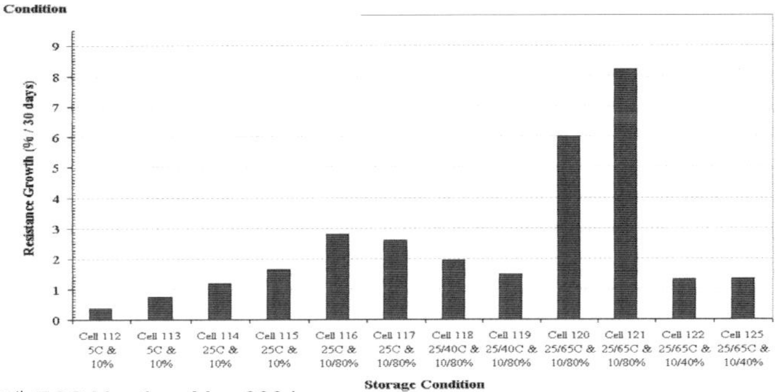


Jeevarajan, Irlbeck, The 205th ECS Meeting, May 2004

## Cycle Life Studies on Li-ion Cells Under Orbiter APU Upgrade Study

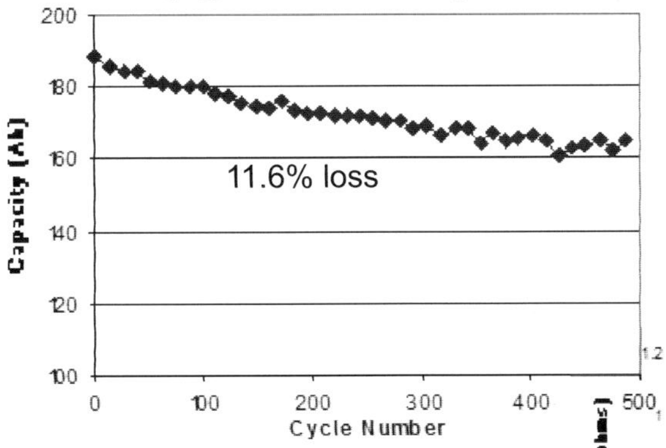


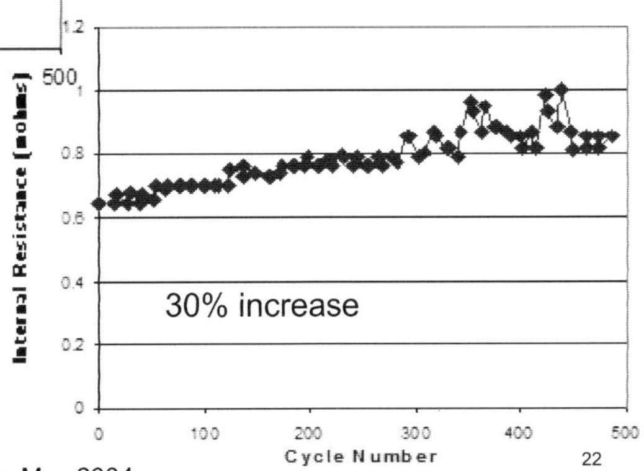


Jeevarajan, Irlbeck, The 205th ECS Meeting, May 2004

22

## Safety after Cycling - Overcharge

1 C current; fresh cell

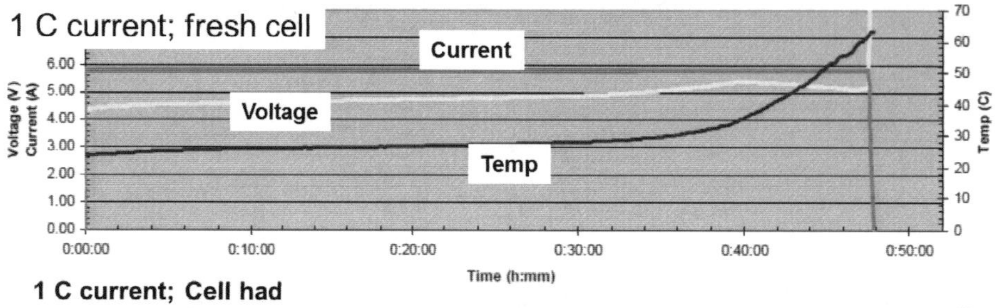


1 C current; Cell had undergone 300 cycles

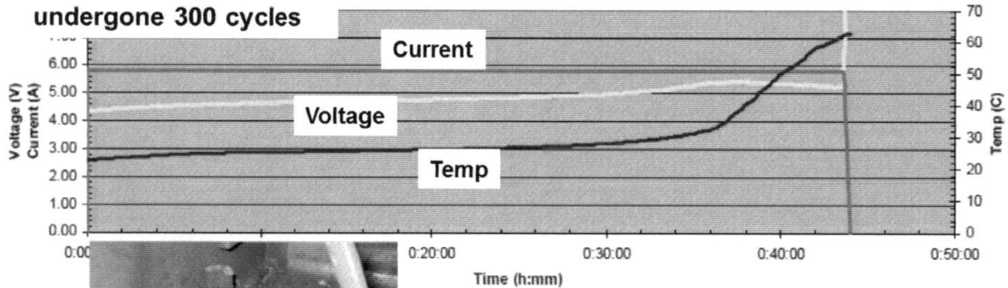

3.1 % Capacity loss after 300 Cycles

J. Jeevarajan, 222nd ECS Meeting, October 2012

## Cycle Life Aging and Simulated Internal Short Tolerance

- **Cells were cycled at 1C rate of charge and discharge for 1000 cycles**
  - **Cells lost capacity between 12 to 25%**
- **Conducted Simulated Internal Short (SIS) tests (Crush test method) - Sample size – 10 cells**
  - **Tolerance to simulated internal shorts increased with higher loss in capacity– no fire or thermal runaway observed even with cells that lost greater than 19% capacity (SIS performed at 100% SOC); cells that lost between 12 to 16% capacity went into thermal runaway (SIS performed at 100 % SOC)**

**Note: All fresh cells at 100 % SOC when subjected to simulated internal short went into thermal runaway**

Jeevarajan, *et.al*, The 2011 NASA Battery Workshop
Jeevarajan, *et.al*, Battery Safety 2012

24

## Current Studies: Test Plan – UL/Texas A&M University

- **Single Cell Studies**
  - Cycle life with continuous cycling at normal voltage range
  - Cycle life with continuous cycling with reduced voltage range (200 mV less from both ends of voltage range)
  - HEV profile at 3 temperatures
  - **Overcharge and external short test on fresh and cycled single cells**

- **Module (3P9S: 9.9 Ah, 33.3 V) Studies**
  - Cycle life with continuous cycling at normal voltage range
  - Cycle life with continuous cycling with reduced voltage range (200 mV less from both ends of voltage range)
  - HEV profile at 3 temperatures (10 °C, 23 °C and 45 °C)
  - **Overcharge and External short test of fresh and cycled modules**

**Destructive Analysis:**

Fresh cells : uncycled, externally shorted cell and overcharged cell

Cycled cells: cycled (cells removed after set number of cycles), externally shorted cycled cells, overcharged cycled cells

Fresh modules : uncycled, cells from shorted and overcharged modules

Cycled modules: cycled, cells from shorted and overcharged modules

25

## Capacity and Internal Resistance Trend with Cycle Life

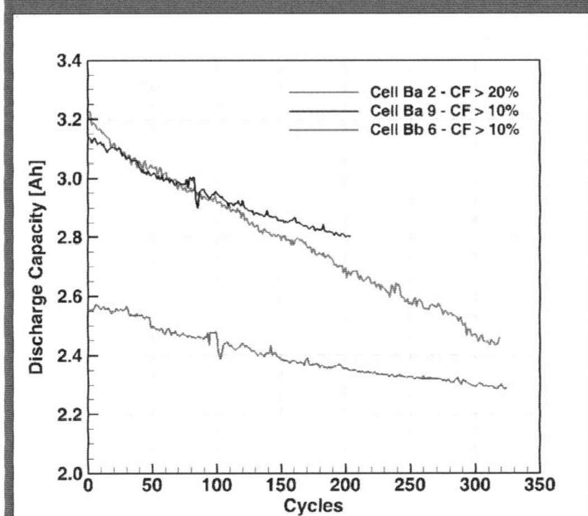


Test Protocols:
Ba: CC/CV with EOCV of 4.2 V
Discharge to 2.7 V
Bb: CC/CV with EOCV of 4.0 V
Discharge to 2.9 V

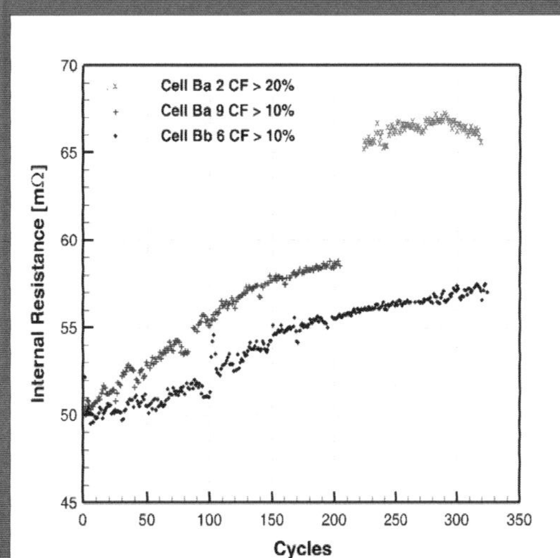

## Challenges with Cell Safety Features

Overcharged cell (fresh cell) with delayed CID activation

Anode showing charring

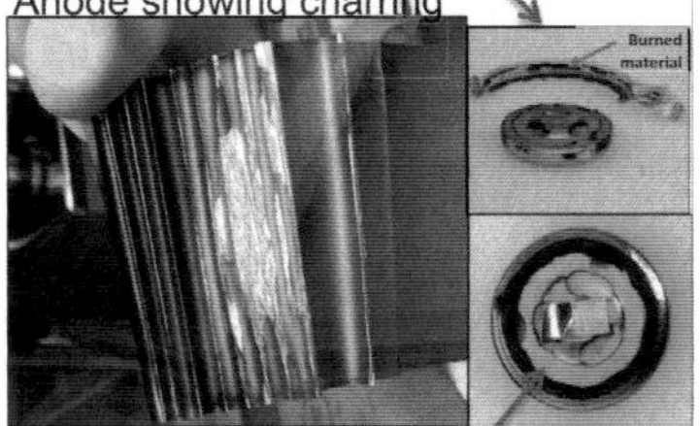


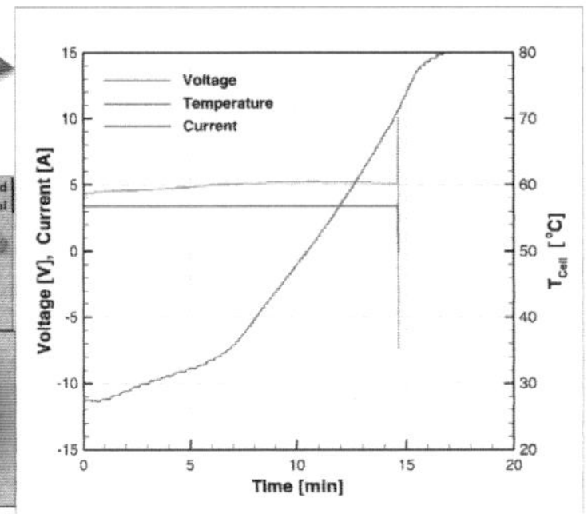


Cell header (underside) showing charring

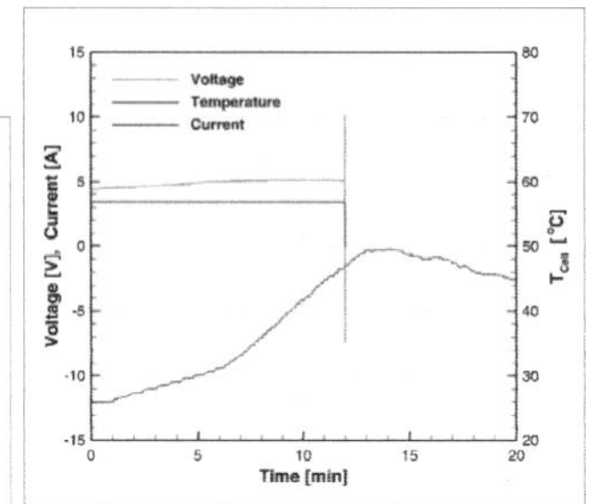


Cell with normal CID activation

Anode of Fresh cell

Cell exhibited PTC activation

Anode surface of fresh (not cycled) externally shorted cell

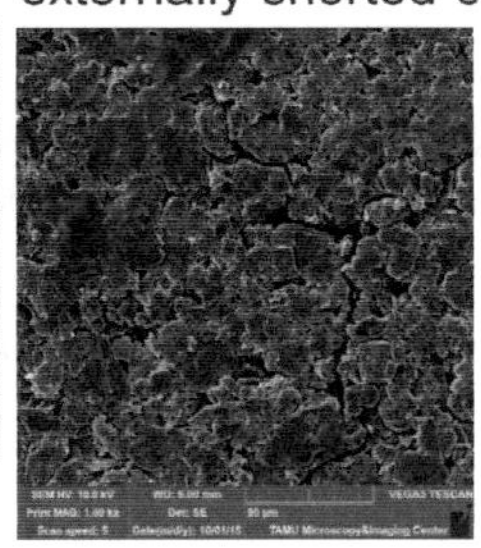

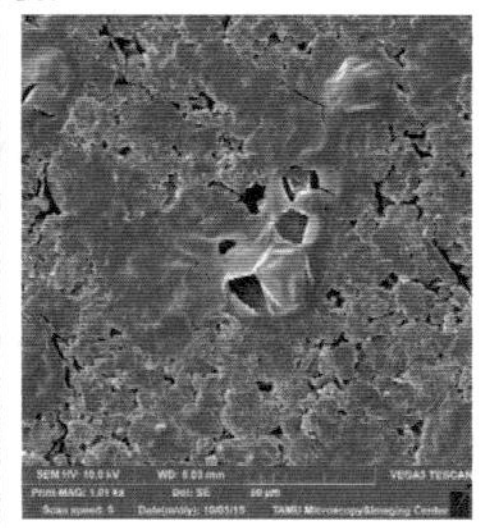

27

## Approaches for Safe Designs for Use and Transportation

- Use **within manufacturer's spec for voltage, current and temperature**

OR

- **Qualify with ample margin to requirements**;

  •Reducing voltage range used by application increases battery life, health and safety

- **Complete characterization of performance and safety** of lithium-ion cells and batteries should be carried out to carve out the baseline (qualification of design)
- **Test stringently** and extensively in the **relevant configuration** and **environment** to understand and **characterize the safety controls and their limitations.**

- Carry out high fidelity thermal analysis and design with best heat dissipation paths

28

## Approaches for Safe Designs for Use and Transportation

Prevent Cell to cell thermal runaway propagation

- Cell to cell spacing; barriers between cells, etc. to prevent cell to cell thermal runaway in the unlikely event of single cell thermal runaway.
- Use of extinguishers and other fire suppression methods installed internal to battery container.

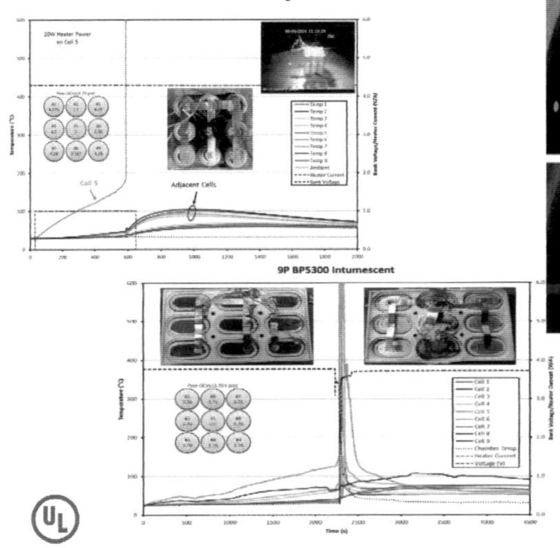


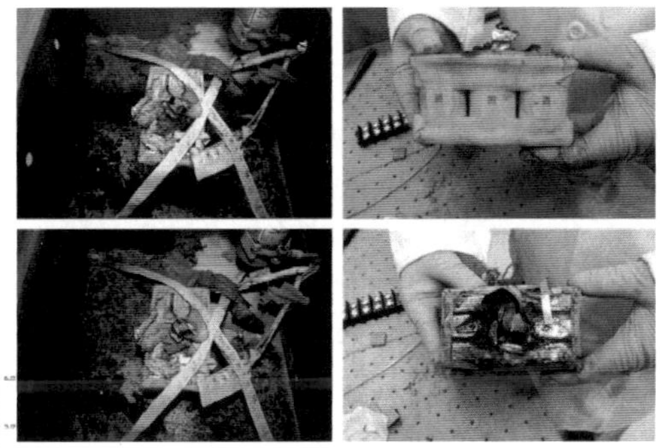

29

## Methods to Determine Module Health Before Second Life

Test and Destructively Analyze sub-module:

- Using **high fidelity thermal analysis** for the battery/module design, the module exposed to worst case thermal deviations should be chosen for testing (as it will not be reused)
- Carry out voltage, capacity and internal resistance/ac impedance tests
- Carry out cell to cell interconnects' integrity tests, complete visual inspection and voltage measurements on each cell /cell bank
- Disassembly of module followed by measurement of cell voltage, capacity of individual cells, internal resistance and ac impedance tests
- Disassembly of cells to study electrodes and electrolyte; three electrode cell studies
- Safety tests at the module and cell level to study any variations in safety characteristics between used and baseline values (inclusion of ARC along with electrical safety tests will add value)

30

## Methods to Determine Module Health Before and During Second Life

For Modules to be reused:

- Visual inspection (should include internal component inspection), voltage, capacity, internal resistance/ac impedance
- Functional checks for proposed new second-use application and environment – run profiles in the relevant environment to confirm that module can perform as required
- Functional checks through complete process of assembly and after assembly into stationary energy storage configuration for utilities.
- Continuous monitoring of health of the cells, modules, battery and system to look for anomalies – allows for early problem detection
- Confirm that charger is suitable for the age of the battery

31

## Summary

- Complete characterization of performance and safety of lithium-ion cells and batteries should be carried out to carve out the baseline (qualification of design)
  - Test stringently and extensively in the **relevant configuration** and environment to understand and characterize the limitations.
- Use **Stringent screening methods** for cells and batteries.
- Being **vigilant of off-nominal behavior and recognizing** this during the life of the battery is a critical part of removing defective cells/batteries from service before they go into a catastrophic failure mode. More important for long term usage batteries.
- Lastly, **usage limits**, **appropriate monitoring and control, cell balancing and thermal design** are key to prevent subtle defects from turning into nucleation sites for larger fault conditions.
- **Analysis and modeling** are ways to lower cost of testing large modules and batteries; however, full scale safety testing should be carried out with the final full scale battery design
- **Test, design, retest, redesign,.....dynamic pattern is required for confirmation of safety**

32

## Acknowledgment

Texas A&M University team for current collaborations with UL:
Prof. Partha Mukherjee
Daniel Robles
Chien-Fan Chen

33

THANK YOU.

# Single-Point Probes for Lithium-Ion Battery Fault Detection and Temperature Monitoring

Corey T. Love and Karen E. Swider-Lyons
Chemistry Division, Naval Research Laboratory, Washington DC

Tatyana Reshetenko, Mathieu Dubarry, Arnaud Devie and Richard Rocheleau
Hawaii Natural Energy Institute, UH-Manoa



## OUTLINE

1. Single-Point Impedance State-of-Health Frequency, $f_{SOH}$
2. Successful Overcharge Detection
3. Application to 4 cells in Series
4. Utility of Current Perturbations
5. Low Temperature Considerations
   - Continuity of $f_{SOH}$
   - Detection of Lithium Plating
6. Intrinsic Link between Impedance & Temperature

***all data collected on same commercial 18650 cell type***

## Success Diagnosing Overcharge Abuse

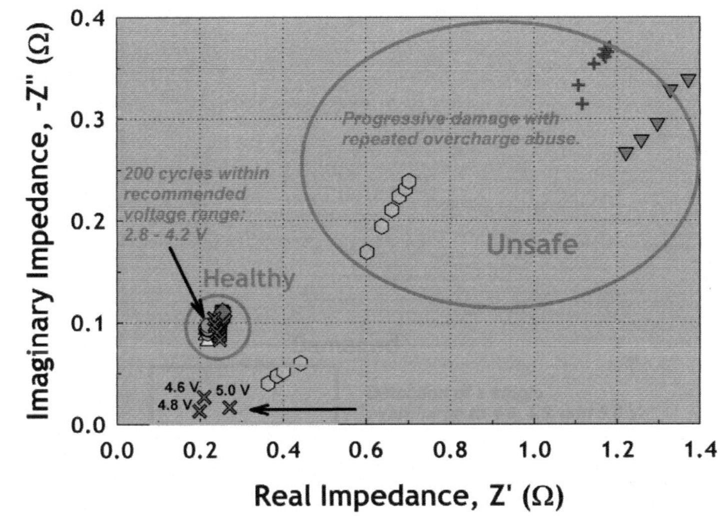


**ADVANTAGES**

- **Does not require collection of full impedance spectrum.**
- **Independent of SOC, can be applied online in real-time.**
- **Minimal instrumentation and hardware needed.**
- **Fast data acquisition and analysis.**
- **Provides binary response: HEALTHY or UNSAFE.**
- **No reliance on equivalent circuit models.**



## Identifying the State-of-Health Frequency Range

Analysis of Phase Angle & Magnitude of Impedance over 0 – 100% SOC

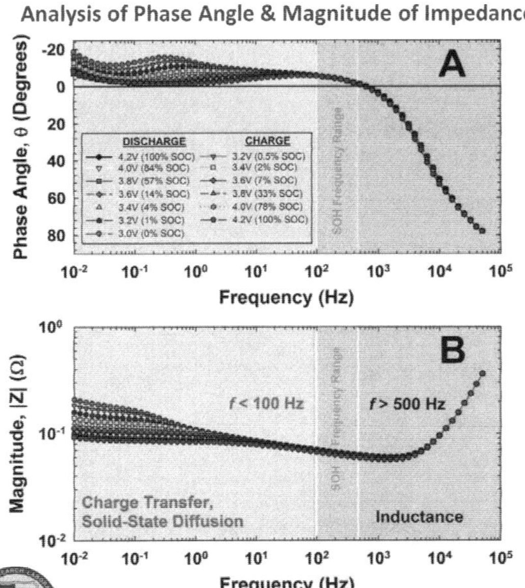


- Eliminate effects of inductance, > 500 Hz
- Reduce variability with SOC, < 100 Hz

**SOH Frequency Range: 100 – 500 Hz**

Standard Deviation of Impedance Response

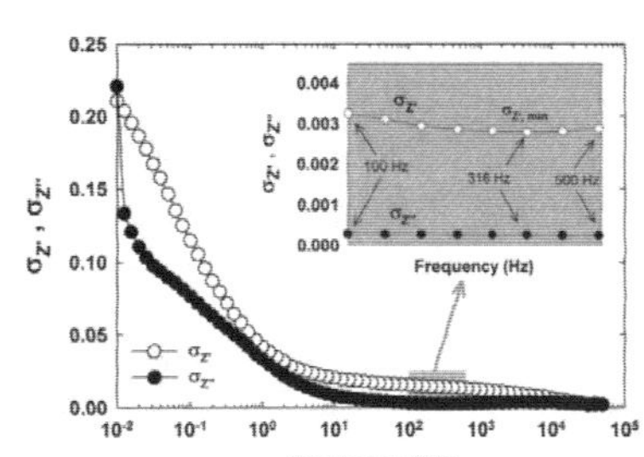


$f_{SOH}$ = **316 Hz**

# Identifying the Single-Point SOH Frequency

**Full EIS Spectrum**
**0 – 100% SOC**

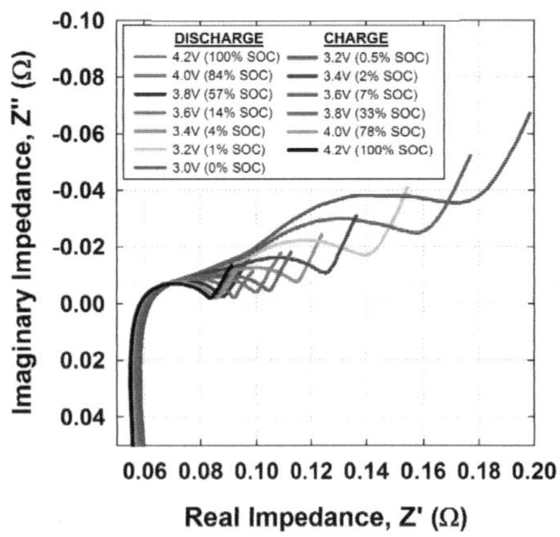


**Average Impedance & Variance**

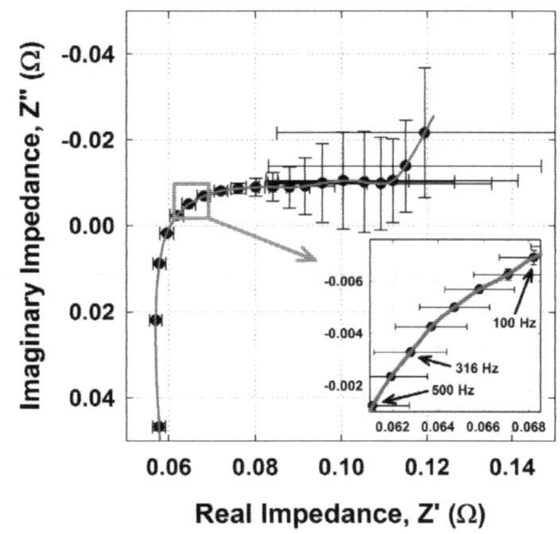




# Probing Cell Impedance at the $f_{SOH}$ = 316 Hz

**Expansion of Healthy Region to 1σ, 2σ and 3σ**

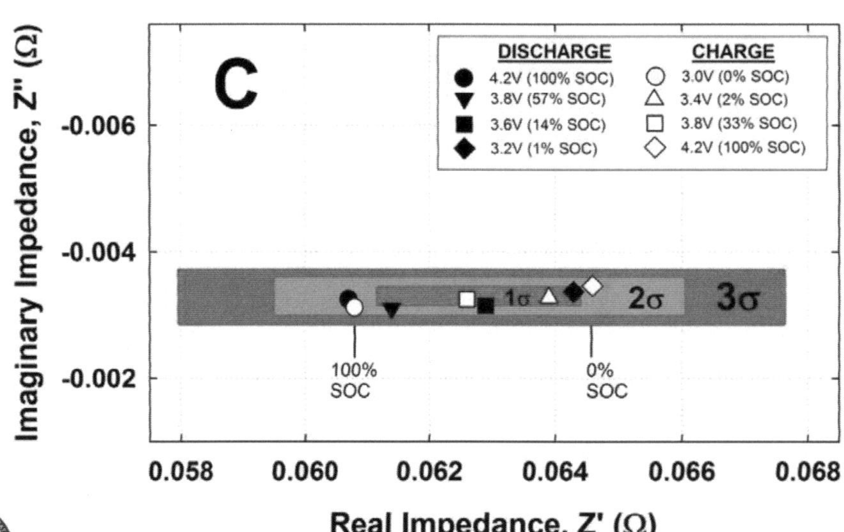

# Diagnosis of Overcharge Abuse

## Single Cell subjected to 5 Overcharge Cycles to 4.56 V

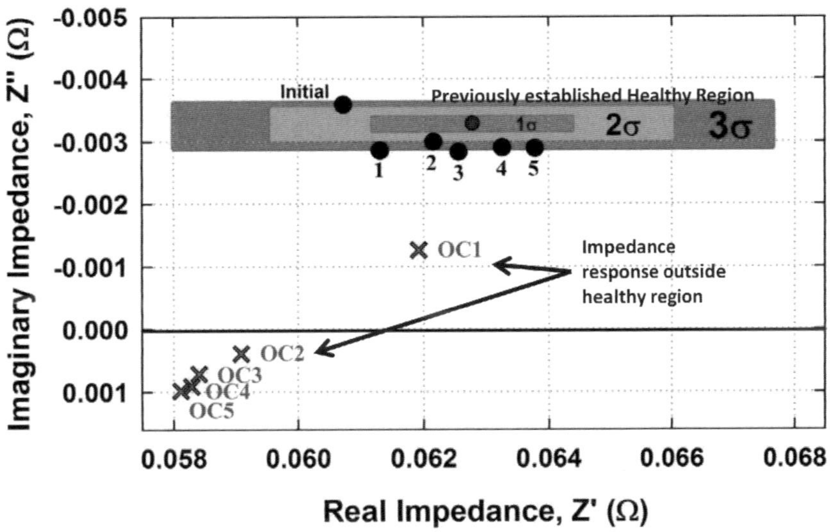




# Utility at Low Temperature, 0°C

**Current Perturbation**

impedance signal-to-noise decreases with increasing amplitude

**Scales with Temperature**

impedance increases as temperature decreases, smallest variance remains at $f_{SOH}$ =316 Hz

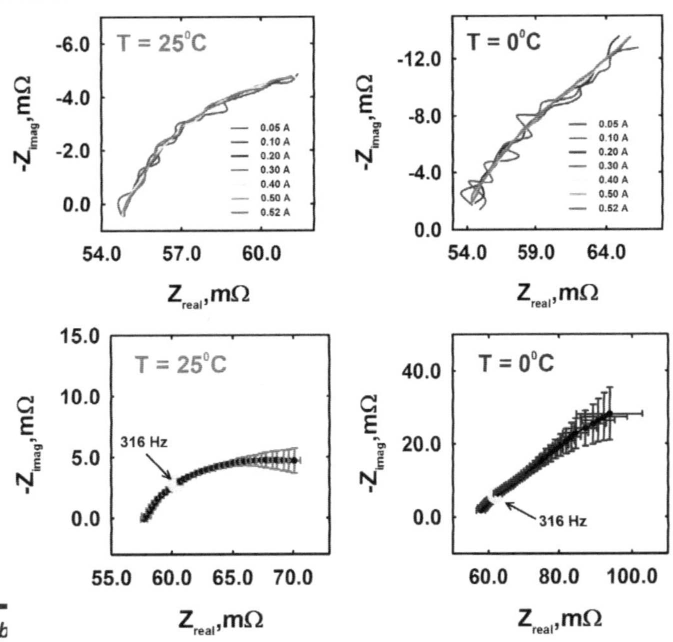

## Empirical Relationships, $Z_{mag}$, $Z_{real}$, $Z_{imag}(T)$

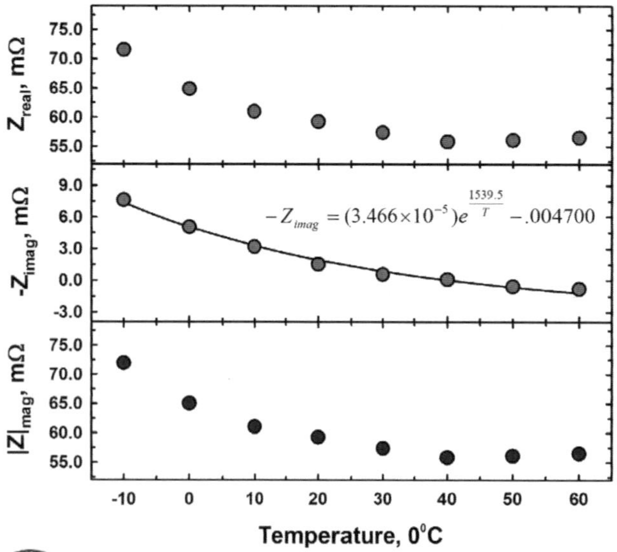


Reversible, Repeatable through 30% Capacity Loss

$Z_{real}$ dominates $Z_{mag}$ ...focus on $Z_{imag}$



## Application to 4S Packs

### Take Advantage of Additive Resistance in Series, No Need for Individual Cell Monitoring

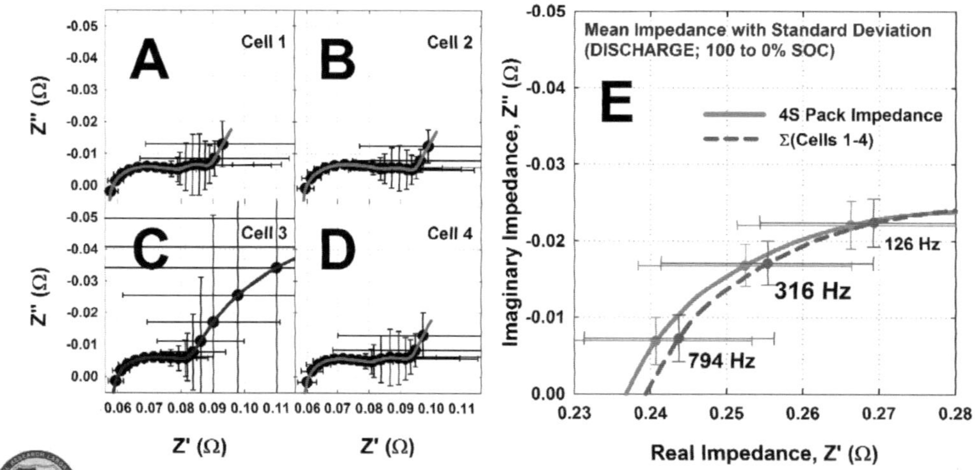

# Diagnose Single Cell Overcharge Abuse within 4S Pack

## Overcharge to Cell 2 Only

II. Cell 2 Overcharged to 4.52 V (125% SOC)
IV. Cell 2 Overcharged to 4.62 V (135% SOC)

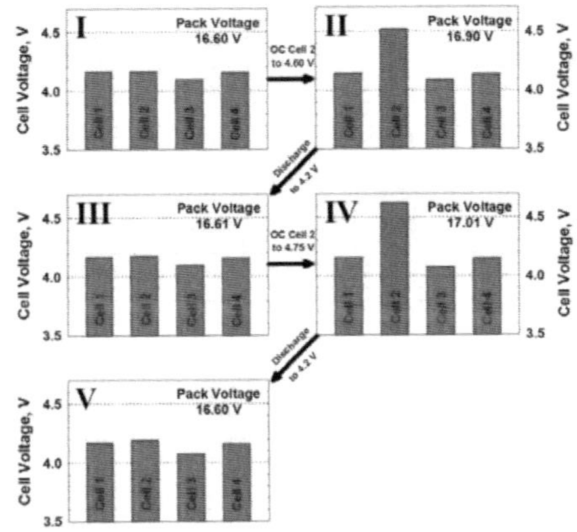


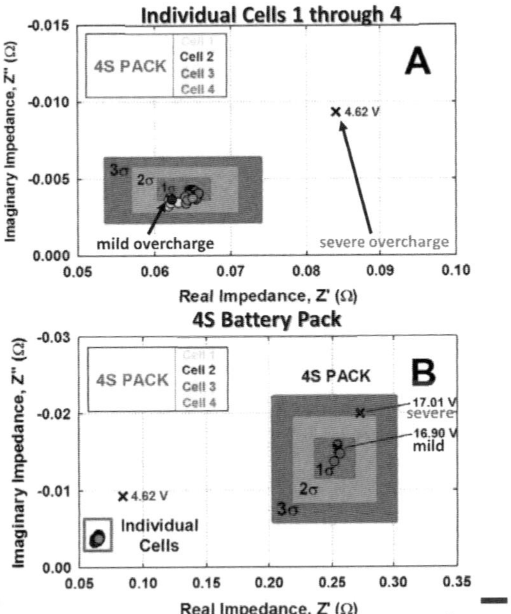


*U.S. Naval Research Laboratory*

*Page 11 / 20*

# Repeated Cycling with Different Cut-off Criteria

## 25°C
### Pack Voltage Cut-off, 11.0 V

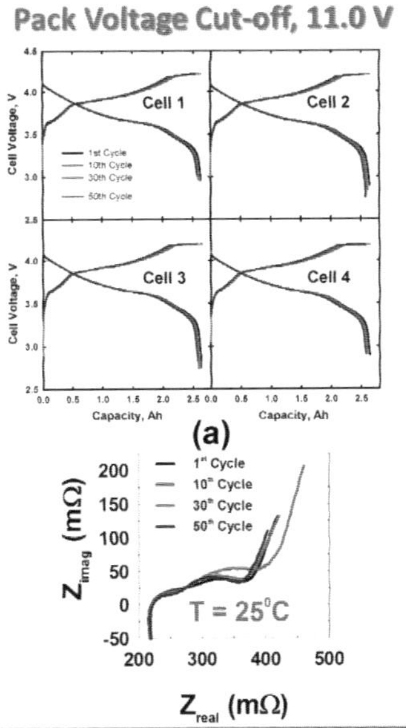


## 0°C
### Cell Voltage Cut-off, 2.75 V

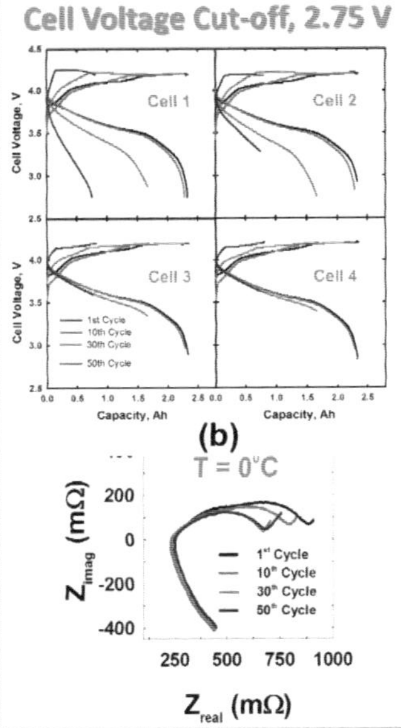


## 0°C
### Pack Voltage Cut-off, 11.0 V

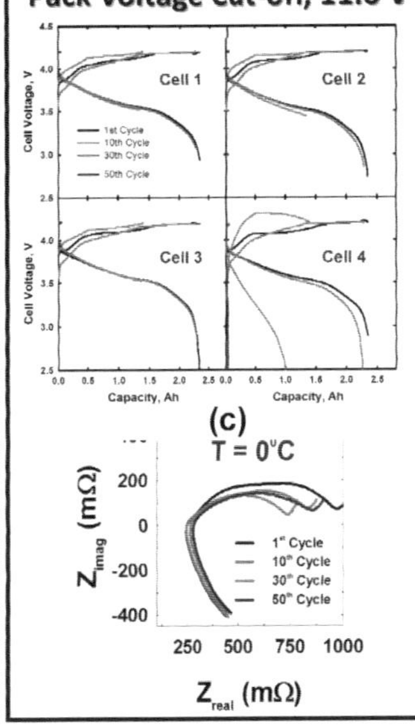

## Capacity Loss with Repeated 0°C Cycling

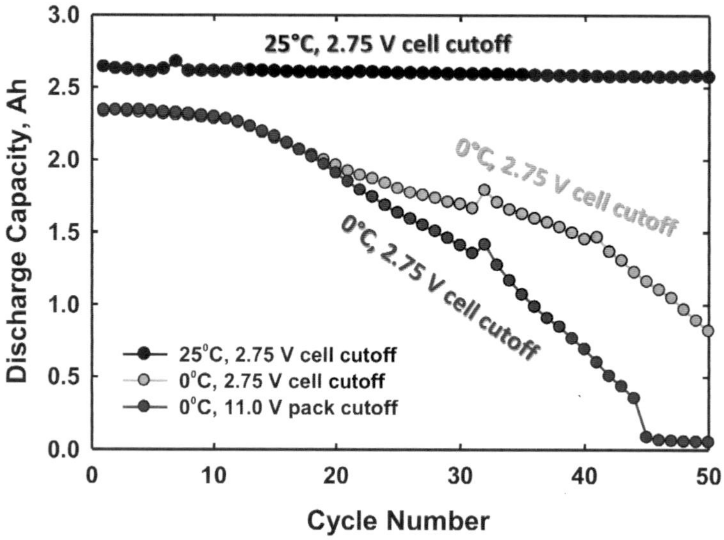




## Temperature Informed SOH Mapping

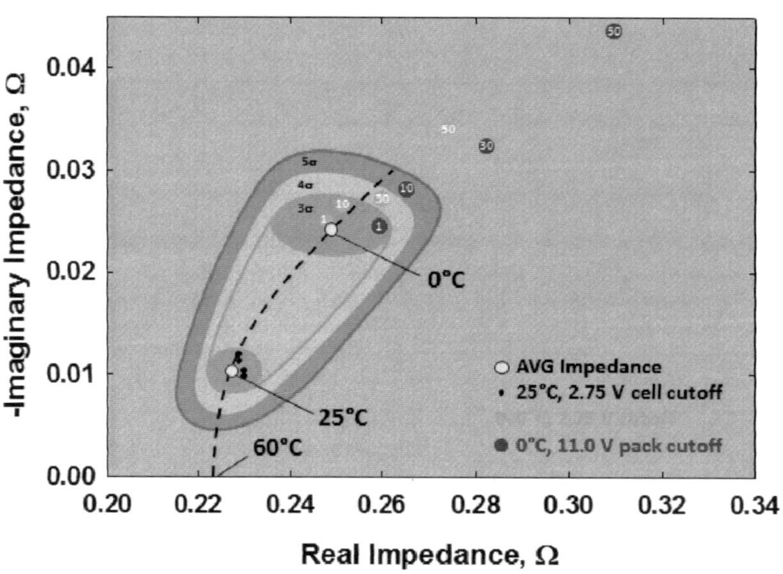

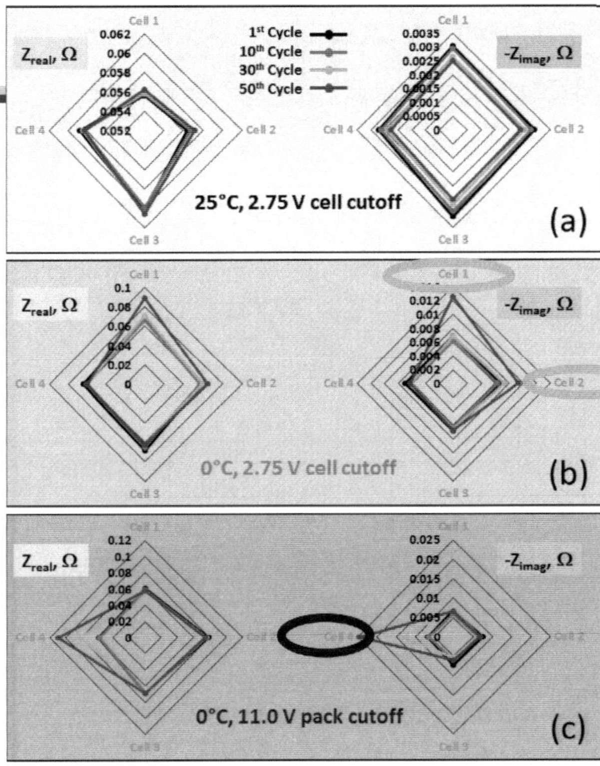


## "Forensic" SOH Mapping at $f_{SOH}$ = 316 Hz

25°C initial impedance mismatch, no propagation after 50 cycles

0°C initial impedance is uniform, rapid increase in $-Z_{imag}$ of Cell #'s 1 & 2 after 30 cycles

**0°C large increase in Cell #4, weak cell driven to overdischarge**

*Page 15 / 20*

# Intrinsic Link Between Impedance & Temperature

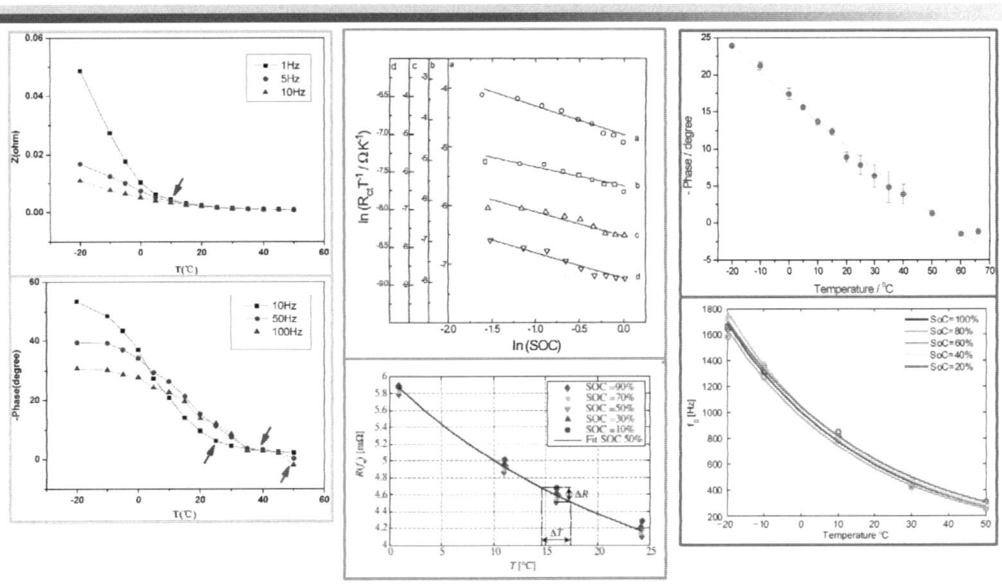


J. G. Zhu, Z. C. Sun, X. Z. Wei, H. F. Dai, J. Power Sources 274 (2015) 990-1004
P. Suresh, A. K. Shukla, N. Munichandraiah, J. Appl. Electrochem. 32 (2002) 267-273
R. Srinivasan, B. G. Carkhuff, M. H. Butler, A. C. Baisden, Electrochim. Acta 56 (2011) 6198-6204
J. P. Schmidt, S. Arnold, A. Loges, D. Werner, T. Wetzel, E. Ivers-Tiffée, J. Power Sources 243 (2013) 110-117
L. . H. J. Raijmakers, D. L. Danilov, J. P. M. van Lammeren, M. J. G. Lammers, P. H. L. Notten, J. Power Sources 247 (2014) 539-544

# Expanding the Operational Limits

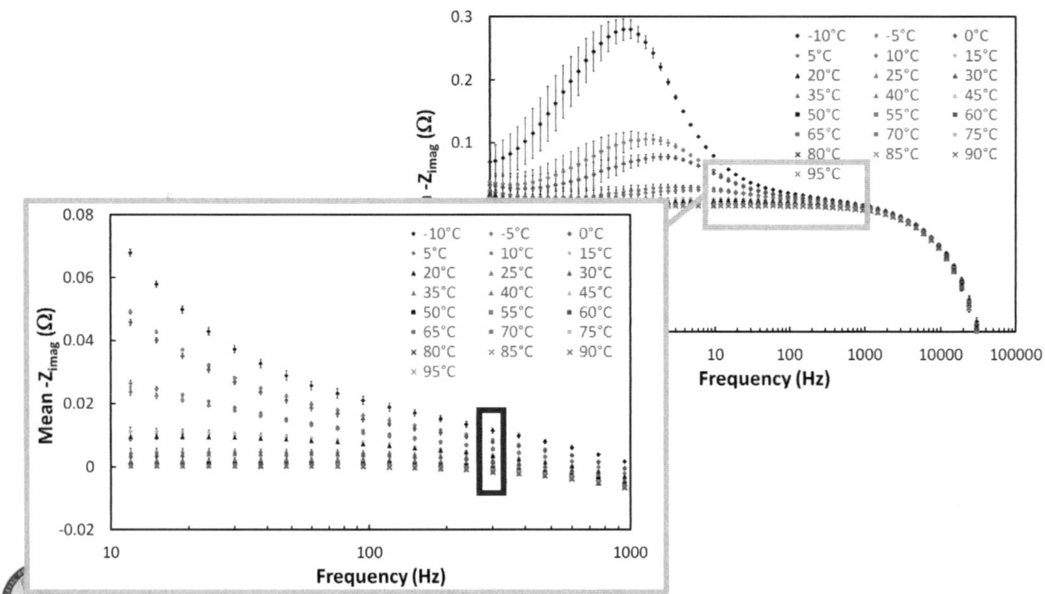




# Near Real-Time Temperature Detection

Dynamic temperature-time program

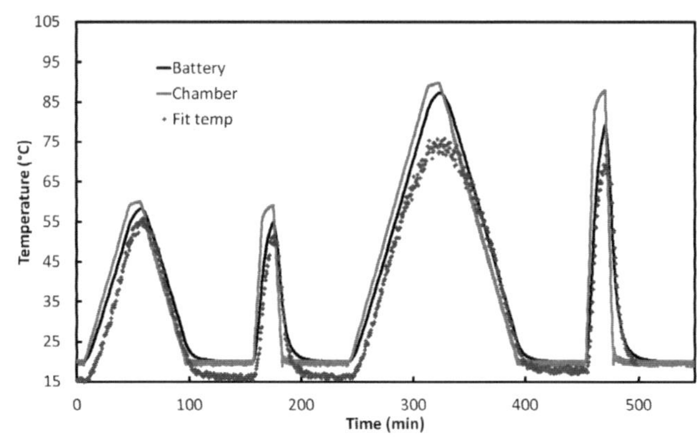


Correlation applied to obtain fit temperatures throughout heating/ cooling steps

Hold until $T_{fit} > 68°C$

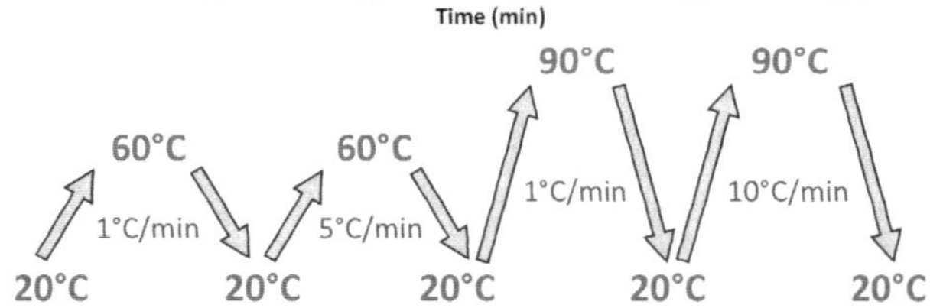

## Summary

1. Utility of, $f_{SOH}$ = 316 Hz holds for overcharge, lithium plating detection
2. Ability to apply diagnostic to 4S strands
3. Empirical relationship with temperature informs SOH with greater specificity
4. Expanded the temperature boundaries of active monitoring



## Acknowledgements

Sponsors:

**Dr. Michele Anderson**

Office of Naval Research Science & Technology

**Mr. Glen Fogg**

Rapid Reaction Technology Office

# Backup Slides

*U.S. Naval Research Laboratory*

# $-Z_{imag}$ vs. $T_{internal}$ Correlation

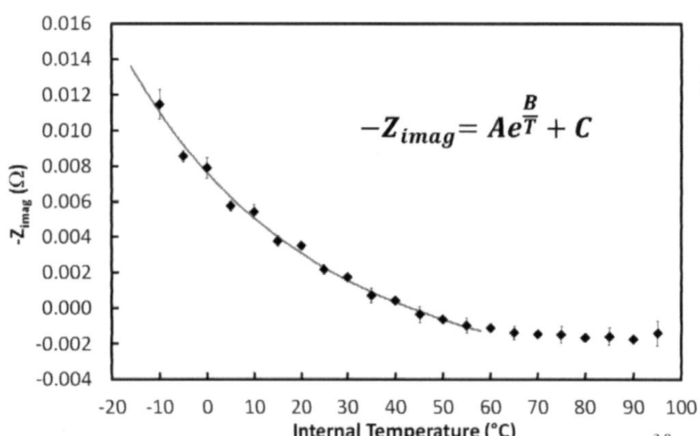


| A | B | C |
|---|---|---|
| 5.041 x $10^{-5}$ | 1539.5 | -0.006553 |

| A (mΩ) | $E_A$ (eV) |
|---|---|
| 0.050 | 0.13 |

$$\ln(-Z_{imag} - C) = \ln A + \frac{B}{T}$$

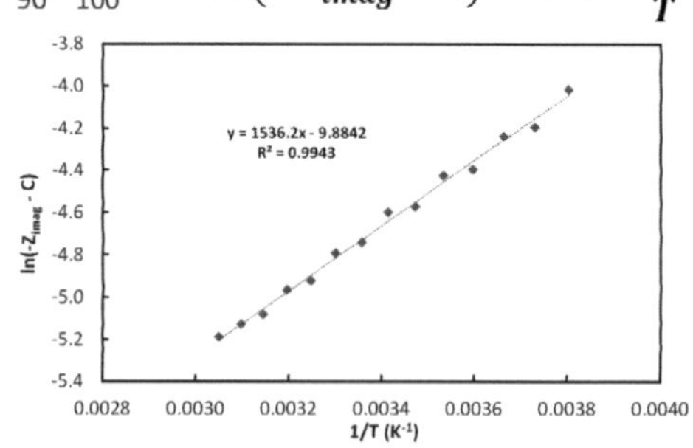


- Virtually instantaneous measurement of internal temperature
- Higher temperatures – possible sensitivity hurdles

*U.S. Naval Research Laboratory*

Morgan
Advanced Materials

# Addressing the Safety Issues Related to Air Transportation of Lithium-Ion Batteries with Effective Engineered Thermal Management Solutions

33rd Annual International Battery Seminar & Exhibit

Richard Clark richard.clark@morganplc.com
Alysha Thomas alysha.thomas@morganplc.com

www.morganadvancedmaterials.com

## Contents

- Overview of Morgan Advanced Materials
- The extent of the challenge
- Overview of legislation
- Materials overview
- Engineered solutions
- Questions

## Morgan Advanced Materials

Founded in England in 1856
Stock symbol on LON: MGAM
2015 revenue: GBP912 million (USD1.3 billion)
2 Divisions and 6 Global Business Units
Thermal Ceramics, including Fire Protection: GBP372 million

Morgan Advanced Materials · Morgan Advanced Materials Confidential Information · 3

## Focus on technically demanding, growth markets

Morgan Advanced Materials · Morgan Advanced Materials Confidential Information · 4

# Morgan Advanced Materials

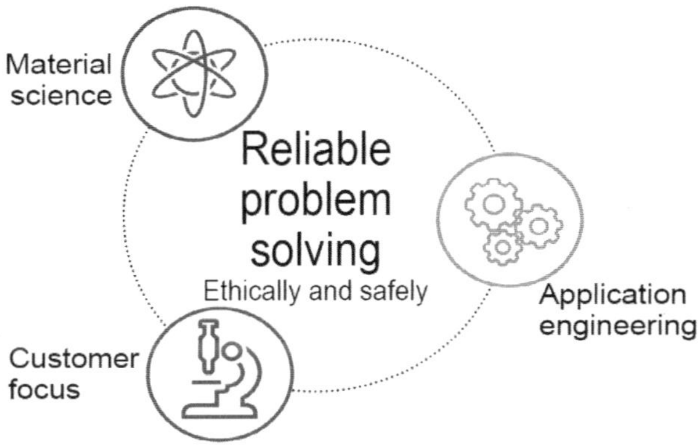


| Global Business Units | |
|---|---|
| Thermal Ceramics | ✓ |
| Molten Metal Systems | |
| Electrical Carbon | ✓ |
| Seals and Bearings | |
| Technical Ceramics | ✓ |
| Composites and Defense Systems | ✓ |

✓ indicates GBU with one or more LIB industry solutions



# Thermal management solutions are wide ranging

FIBER
- Bulk
- Blanket
- Modules
- Converted
- Paper

IFB & FIREBRICK
- JM, K & TJM Grades

MONOLITHICS
- Dense Monolithics
- LW Monolithics
- Special Shapes

SPECIALTIES
- Heat Shields
- Graphite Insulation
- Engineered Insulation
- Microporous Insulation

Morgan engineers and installs high performance insulation in operating environments from -70°C to 3000°C

# Morgan's products on an aircraft

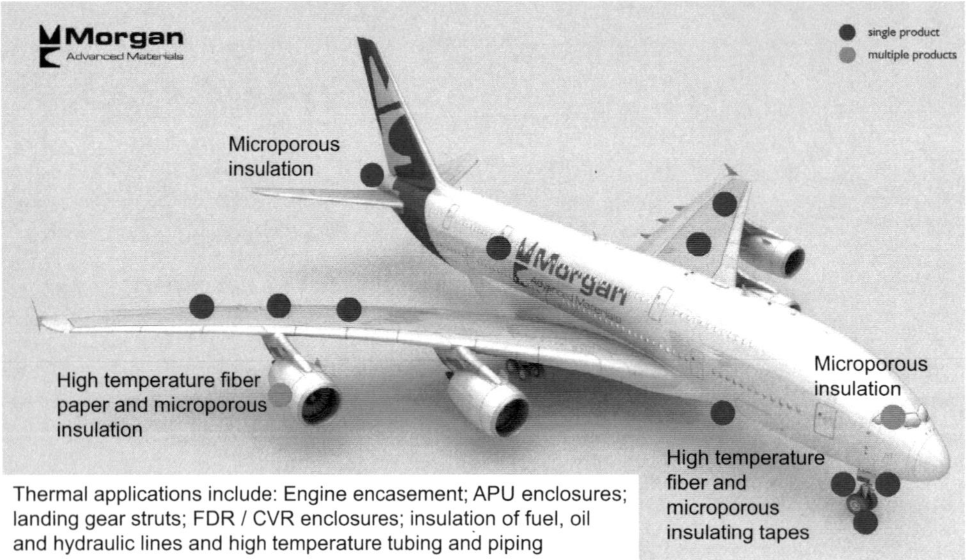




# Contents

- Overview of Morgan Advanced Materials
- **The extent of the challenge**
- Overview of legislation
- Materials overview
- Engineered solutions
- Questions

# Batteries on an aircraft

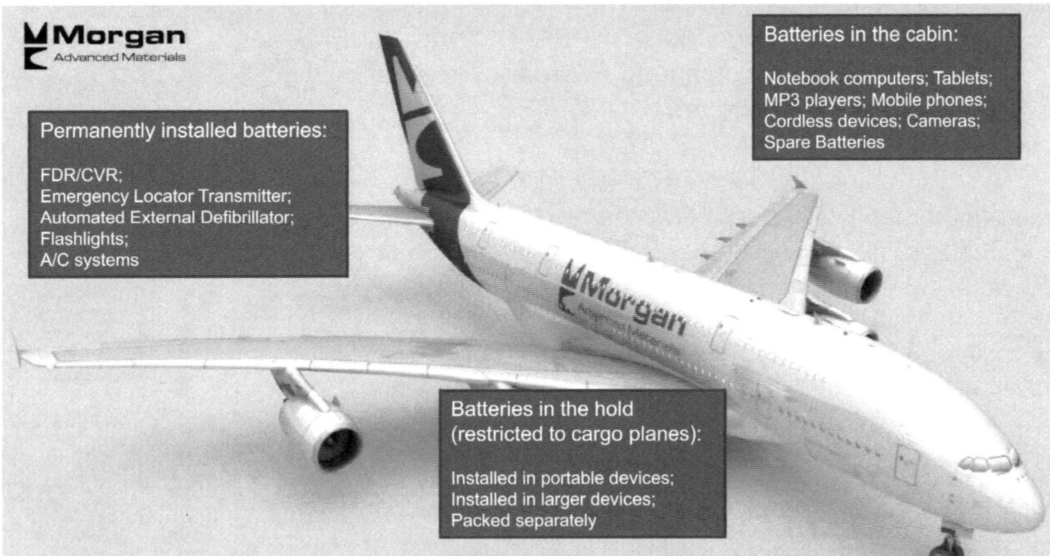


Derived from "Risks related to Lithium Batteries," C. Bezard, 18th Flight Safety Conference, March 2012



# Growing number of incidents

Between March 20, 1991 and January 15, 2016, 171 air/airport incidents involving batteries carried as cargo or baggage have been recorded by FAA

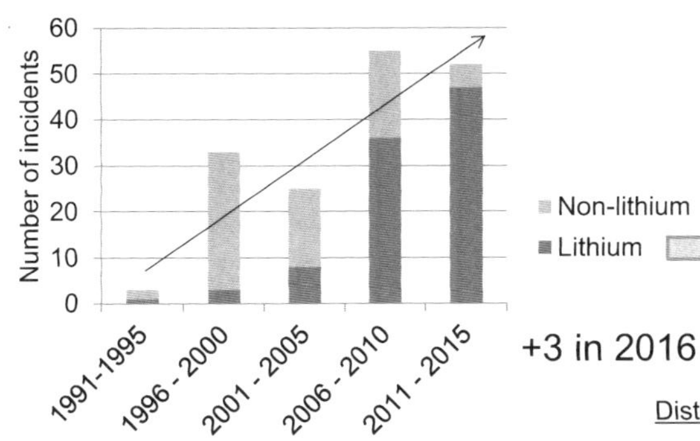


+3 in 2016 YTD

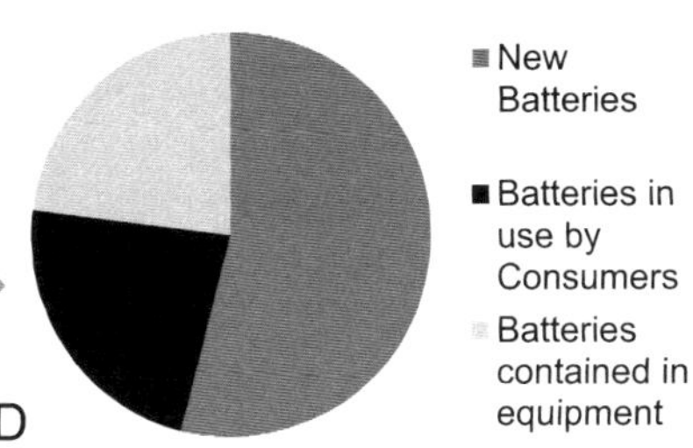


Distribution of Lithium battery incidents in air transport

Data source: http://www.faa.gov/about/office_org/headquarters_offices/ash/ash_programs/hazmat/aircarrier_info/media/battery_incident_chart.pdf

Derived from presentation "Navigating the Regulatory Jungle": Hwang and Leary ,USDOT, Baltimore, 2015

## Catastrophic events attributed to lithium-ion batteries

- Fire erupted in a cargo plane (UPS – DC8) that landed in Philadelphia on February 7, 2006 – airplane destroyed
- A cargo plane (UPS – B747) with 81,000 lithium batteries caught fire and crashed after it left Dubai on September 3, 2010 – airplane destroyed, 2 crew members killed
- A cargo jet (Asiana – B747) crashed into the East China Sea on July 28, 2011, after the crew reported a fire on board – airplane destroyed, 2 crew members killed

Information and pictures from: DOT/FAA/TC-13/2; https://aviation-safety.net/airlinesafety; "Fire Hazards of Lithium-Ion Batteries" R. E. Lyon et al., International Aircraft Systems Fire Protection Working Group Meeting, October 21 to 22, 2015; "Fire Suppression in a Class E Cargo Compartment" Dhaval Dadia, ICAO Meeting, October 30-31, 2013

Morgan Advanced Materials

Morgan Advanced Materials Confidential Information 11

## Contents

- Overview of Morgan Advanced Materials
- The extent of the challenge
- Overview of legislation
- Materials overview
- Engineered solutions
- Questions

Morgan Advanced Materials

Morgan Advanced Materials Confidential Information 12

## Legislation history – Primary Lithium (UN 3090)

- April 28, 1999 Fire at LAX -120,000 lithium primary cells led to issue of FAA DGAB-00-02 on September 7, 2000
- Subsequent studies showed that lithium metal primary cell fires cannot be suppressed by halon
- Banned on passenger aircraft by FAA in 2005 – details in DOT PHMSA HM-224C (April 2, 2002) and HM-224E (December 15, 2004)
- Banned on passenger aircraft by ICAO starting January 1, 2015 – details in ICAO Technical Instruction for the Safe Transport of Dangerous Goods by Air and the 56th Edition of the IATA Dangerous Goods Regulations (DGR)

http://phmsa.dot.gov/hazmat/regs/rulemaking/archive/hm-new

https://www.iata.org/whatwedo/cargo/dgr/Documents/lithium-battery-guidance-document-2015-en.pdf

Morgan Advanced Materials

Morgan Advanced Materials Confidential Information 13

## Legislation history – Lithium-Ion (UN 3480)

- Lithium batteries (both UN 3090 and UN 3480) are currently classified as Class 9 hazardous materials in Title 49 CFR, Hazardous Materials Regulations (HMR) and the ICAO Technical Instructions
- FAA (SAFO 16001) supports recommendations that before operators engage in the transport of lithium ion batteries as cargo on passenger aircraft, or lithium ion or lithium metal batteries on cargo aircraft, they should be aware that ICAO and major airframe manufacturers (Boeing and Airbus) have recommended that operators perform safety risk assessments in order to establish whether, or how, they can manage the risk associated with the transport of these batteries
- Transportation by air of lithium-ion batteries is not currently banned in either cargo or passenger aircraft. However, this is about to change.

http://www.faa.gov/other_visit/aviation_industry/airline_operators/airline_safety/safo/all_safos/media/2016/SAFO16001.pdf

Morgan Advanced Materials

Morgan Advanced Materials Confidential Information 14

## New legislation with effect from April 1, 2016

- On February 22, 2016, the ICAO Council passed a prohibition on lithium ion batteries (UN 3480) on passenger aircraft as cargo that goes into effect on April 1, 2016.
- ICAO "Technical Instructions for the Safe Transport of Dangerous Goods by Air" 2015-2016 (Doc 9284) incorporating corrigenda and addenda up to and including addendum number 4 (February 23, 2016) include the above and other new requirements which include:
  - Lithium ion cells and batteries must be offered for transport at a state of charge (SoC) not exceeding 30% of their rated design capacity
  - Details related to acceptable packing and labeling

http://www.icao.int/Newsroom/Pages/ICAO-Council-Prohibits-Lithium-Ion-Cargo-Shipments-on-Passenger-Aircraft.aspx



## Contents

- Overview of Morgan Advanced Materials
- The extent of the challenge
- Overview of legislation
- Materials overview
- Engineered solutions
- Questions

# Types of High Temperature Insulation Fibers

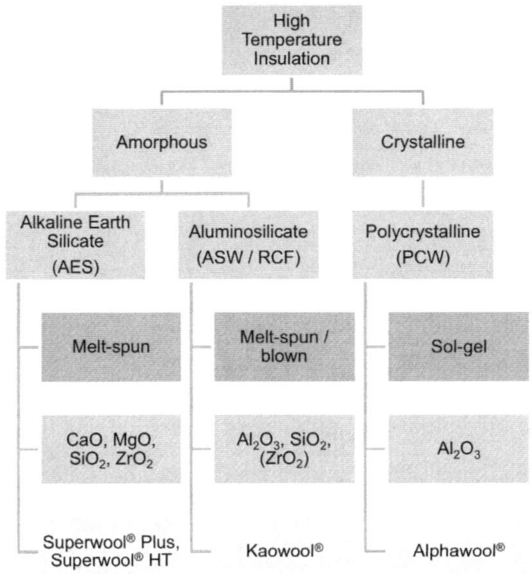


ASW/RCF invented 1942
PCW invented 1969
AES invented 1986
(Microporous invented 1958)

Key Advantages of Superwool® over RCF:

- Low bio-persistence
- Low shrinkage up to classification temperature
- Low thermal conductivity



# What is microporous insulation?

- A low density powder comprised of heat resistant particles and high temperature fibers compressed to form a microporous structure
- The most efficient insulation commercially available for a wide temperature range, engineered to control all three modes of heat transfer – convection, conduction and radiation
- Inorganic material, recyclable and safe from an H&S perspective

## Material optimization / application engineering

Best

Insulating Value

**Key Selection Criteria**

- Application temperature
- Chemical Resistance
- Environmental Factors
- Installed weight and thickness
- Installation requirements
- Initial and total cost of ownership

Structural Aspects — Strongest



## Thermal conductivity comparison

Thermal conductivity comparison of WDS microporous insulation products vs. other types of insulating products

- Perlite (granular)
- Mineral Wool (100 Kg/m³)
- RCF blanket (130 Kg/m³)
- Low shot AES blanket (130 Kg/m³)
- LD Calcium Silicate
- AES fiber board (300 Kg/m³)
- Opacified LD Calcium Silicate
- Aerogel blanket
- WDS® insulation

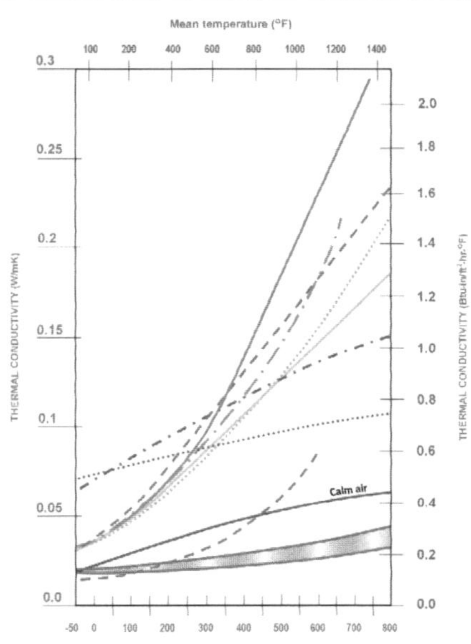

## Contents

- Overview of Morgan Advanced Materials
- The extent of the challenge
- Overview of legislation
- Materials overview
- Engineered solutions
- Questions



## Aviation industry safety challenges with Lithium Ion

6 billion Lithium Ion cells were manufactured in 2015 – 30% were transported by air

| Situation | Immediate consequence | Ultimate consequence |
|---|---|---|
| Thermal runaway in packed Lithium Ion cell | Transmission to large number of adjacent cells | Fire / explosion / pressure wave |
| Thermal runaway in Lithium Ion cell in use | (Transmission to adjacent cells) | Fire / explosion / pressure wave |
| Fire external to package | Elevated cell temperatures start thermal runaway | Fire / explosion / pressure wave |

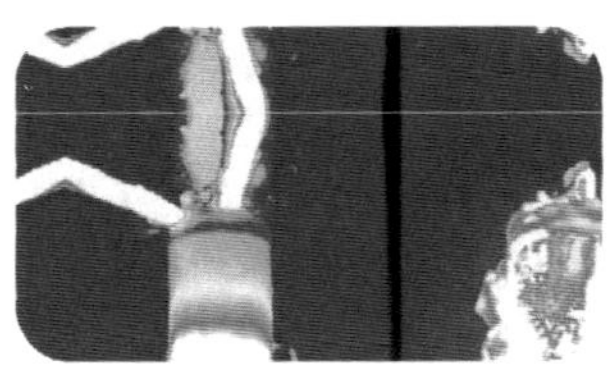

← < 1 second →

Thermal Runaway
- Auto-accelerating heat generation
- Rapid temperature increase
- Expulsion of flammable gases and liquids

"Fire Hazards of Lithium-Ion Batteries" R. E. Lyon et al., International Aircraft Systems Fire Protection Working Group Meeting October 21 to 22, 2015

Many causes of thermal runaway – electrical, mechanical, thermal

## Additional challenges in all applications

- Ideally would employ chemistries and cell structures to prevent thermal runaway from starting, but increasing energy densities make this increasingly difficult, especially in applications where space is critical
- Once thermal runaway in one cell has started then the challenge is to prevent propagation from cell-to-cell, from module-to-module and from pack to surroundings – this applies in electronics, electric vehicle, energy storage and other applications
- In the case of aviation, the key is to extend the time needed for remediation measures to be employed where possible and to land the plane safely

Images from: "Fire Hazards of Lithium-Ion Batteries" R. E. Lyon et al., International Aircraft Systems Fire Protection Working Group Meeting October 21 to 22, 2015 and "Fire Suppression in a Class E Cargo Compartment" Dhaval Dadia, ICAO Meeting, October 30-31, 2013

Morgan Advanced Materials

Morgan Advanced Materials Confidential Information 23

## Morgan Advanced Materials expertise

- Morgan's understanding of the materials science required to prevent the spread of thermal energy is demonstrated in our supply of the material surrounding FDR and CVR products which is designed to survive arduous mechanical and thermal conditions
- We are also the trusted supplier of materials for surrounding living quarters in offshore oil rigs, specifically designed to protect the structure to allow additional time for operators to escape from a fire

Living Quarter Module for Gudrun Offshore Platform

10,000 m² FireMaster Marine Plus Blanket installed in 2011 for one hour fire protection of external and internal structure

Morgan Advanced Materials

Morgan Advanced Materials Confidential Information 24

## New product recently deployed by one major airline

A composite product incorporating FireMaster® Marine Plus Blanket

FireMaster® Battery Bag

- PED containment bag for emergency storage of defective personal electronic devices to avoid expansion of fire in case of thermal runaway of Li Ion batteries.

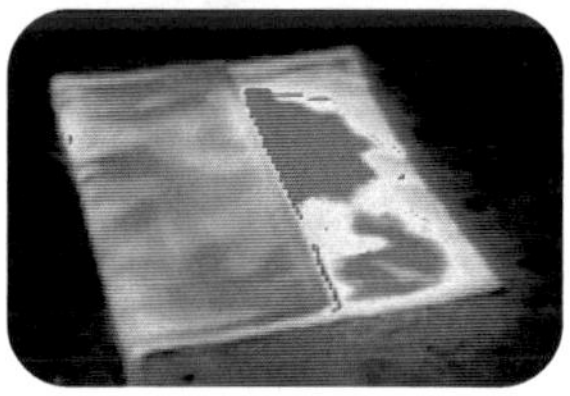

Pictures and videos created by and used with permission of Lufthansa/Germanwings

Morgan Advanced Materials

Morgan Advanced Materials Confidential Information

25

## Moving from mitigation to prevention

- To address the EV and ESS market opportunities for LIB requires:
  - high energy density
  - low cost
  - uncompromising safety
- > 500 Wh/L
- < $125/kWh including electronics
- Intrinsic safety

- A visionary team led by Christina Lampe-Onnerud has conceived of a way to address these requirements
- Morgan is proud to be partnering with Cadenza Innovation in developing a novel engineered housing solution incorporated within the Cadenza solution

Cadenza Innovation

Morgan Advanced Materials

Morgan Advanced Materials Confidential Information

26

## Cadenza Technology Platform: Multi-Jellyroll Architecture

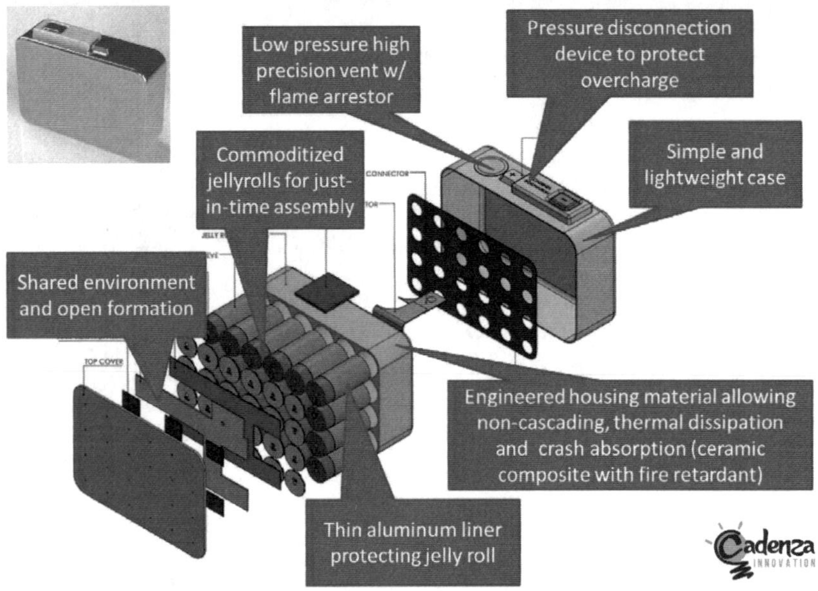


Morgan Advanced Materials

Morgan Advanced Materials Confidential Information 27

## Conclusions

- Current chemistries and cell structures of the vast majority of lithium ion batteries cannot protect against the possibility of thermal runaway in certain conditions
- It continues to be necessary for large quantities of lithium ion batteries to be transported by air
- Although legislation continues to improve the requirements for safe air transport of lithium-ion batteries, there is still a significant and urgent need to provide additional protection for the passengers, crew and aircraft
- As the leading global manufacturer of thermal fire protection systems, Morgan Advanced Materials is working with the air transportation industry to develop solutions to address this need

FIREMASTER
www.morganthermalceramics.com

For more information, please contact:
richard.clark@morganplc.com, +1-216-702-0218
alysha.thomas@morganplc.com, +1-706-414-3075

Morgan Advanced Materials

Morgan Advanced Materials Confidential Information 28

Morgan
Advanced Materials

# Questions?

www.morganadvancedmaterials.com

29

AEROSPACE
Assuring Space Mission Success

# Effects of Cell Reversal on Li-ion Batteries

E. Joseph Nemanick, David Wang, James Matsumoto, Jasen Ross, and Neil Ives

Energy Technology Department

March 24, 2016

The Aerospace Corporation
Physical Sciences Laboratory
Electronics and Photonics Laboratory
Energy Technology Department



## Reversal in Li-ion Cells

- Cell reversal is the condition of the anode electrochemical potential rising above that of the cathode, resulting in a negative voltage measured at the cell level.
- There are two primary reactions that occur at the anode at high potentials which increase cell impedance:
    - Oxidation of copper current collector
    - Oxidization of carbonate electrolytes to $CO_2$
- At the cathode, the reducing potential can lead to the electrodepostion of copper to form dendrites. These copper dendrites pose a shorting risk.
- Cell reversal can be caused by poorly matched cells, a failure of the battery management electronics, or a defective cell in a pack.

## Generating reversal in Li-ion Cells

- Drive reversal in Panasonic NCR18650B cells
  - Testing in a parallel pack comprised of 5 cells in parallel (Pack capacity = 5 * 3.2 Ahr = 16 Ah).
    - Cells are matched before pack assembly.
    - Cell level measurements with matched shunts.

- Test to simulate failure of charge control electronics in GEO eclipse season operation (60% DOD).

- Testing procedure:
  - 1.2 hour discharge at rate of 8 A, C/2.
  - 15 minute open circuit.
  - Charge at C/10 (1.6A). Total time is set to recharge the selected capacity to 90% nominal capacity.

  - Repeat the charge/discharge cycle until cells have been voltage reversed and an internal short has developed.



## Voltage profile of virtual cells driven into reversal

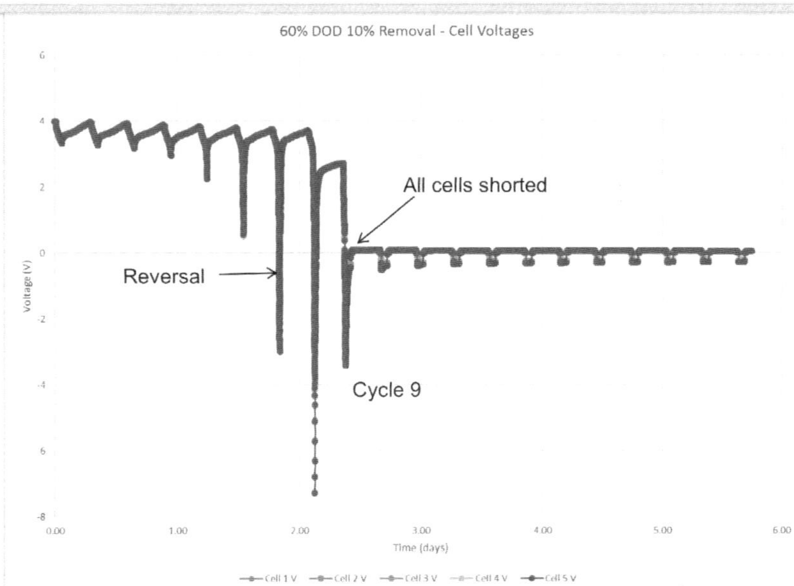


Cells failed at expected cycle, 5, a 60% DOD of cells that started at 60% SOC.
All cells developed shorts after 9 cycles.

## Cell temperatures

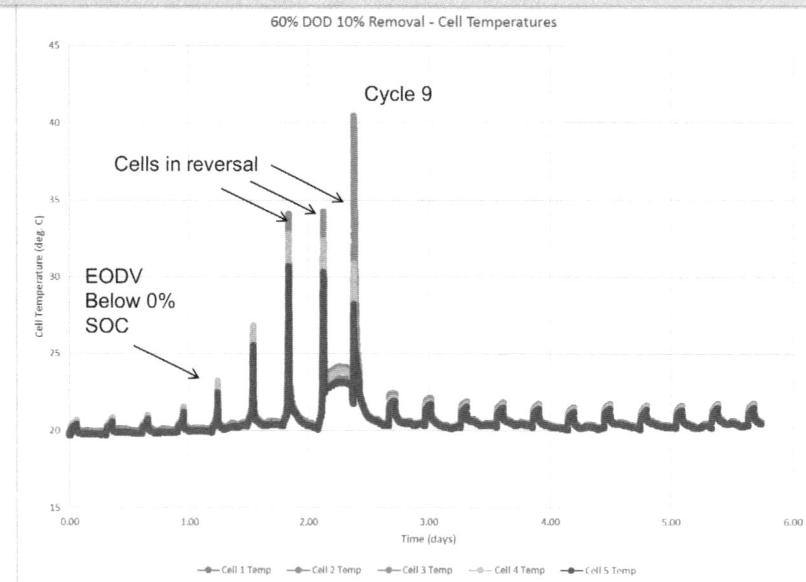


Cell temperatures begin to rise as discharge approaches 0% SOC, and spike to 40+ C in reversal.



## Cell currents

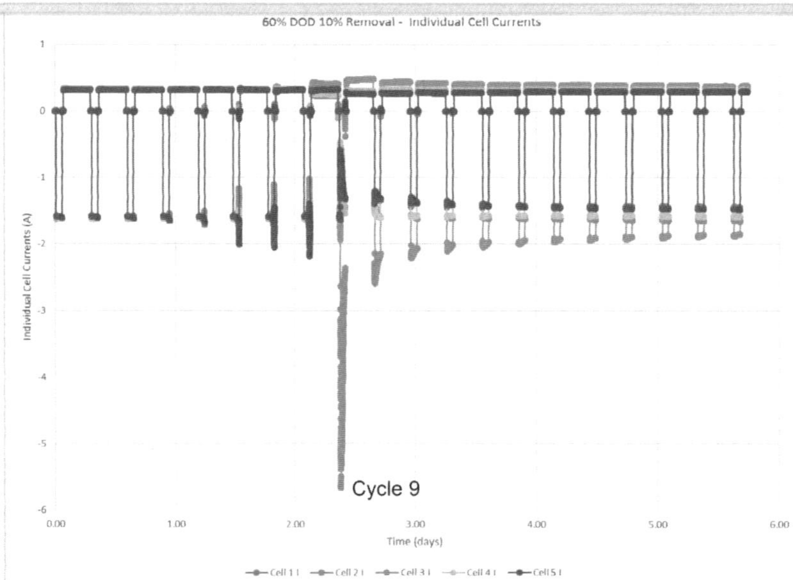


Cell 3, the highest temperature cell in reversal, carried majority of total current at peak. Likely the first to short. However, other cells quickly shorted.

## Electrical testing conclusions

- Cells fell below 0% SOC after 5 cycles. Nominal initial SOC went from 100% SOC on first cycle to 60% SOC at start of cycle 5.
- Cells still cycled for 3-4 more cycles before shorting. Cells did not immediately fail after going into reversal. Passing back into normal operation after reversal should corrode Cu deposited on the cathode and redeposit a portion on the anode, though not on current collector.
- The initial short occurred in cell 3, as shown by the current sharing. However, the other cells quickly shorted as well, creating a more equal current sharing. A short in one cell did not prevent other cells from shorting.



## Non destructive detection of copper plating by CT X-ray

Reversed

Control

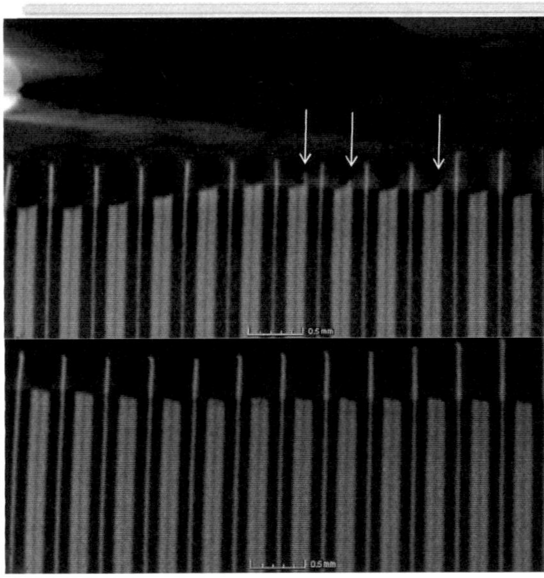


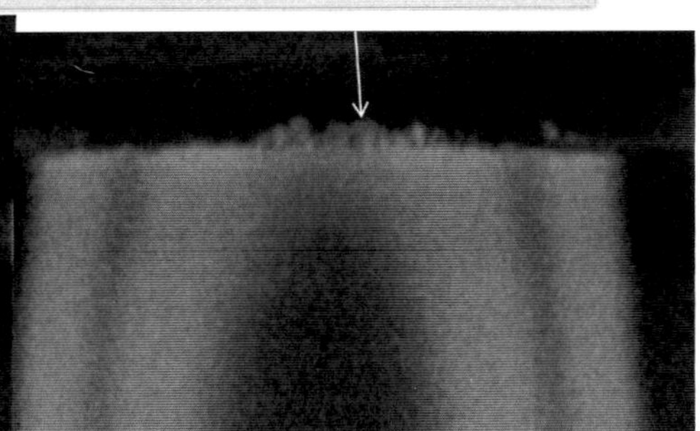


Copper dendrites are observed adjacent to the separator and rise upward towards the copper gradient.

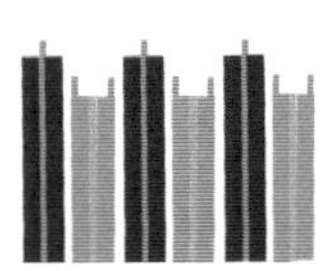

CT X-ray scans could not detect Cu plating within the cell stack. However, dendrites at the cathode top edge were very visible.

## CT X-ray of reversed cells

Horizontal plane slice

Vertical plane slice

Copper dendrites rising up from edge of NCA

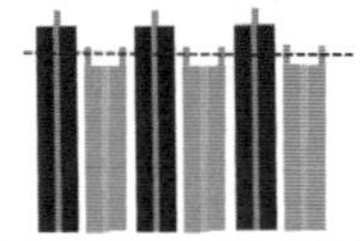

Copper transport from the negative current collector should deposit the majority of Cu within the stack. However, the edge overlap of the anode also causes Cu to deposit on the edge of the positive electrode.

While the resolution of these CT was better than 20 μm, far less is necessary to detect copper at the cathode edge.

ejoseph.nemanick@aero.org EPL/Energy Technology Group

AEROSPACE

9

## Cell destructive physical analysis

Cell leakage and copper colored electrolyte.

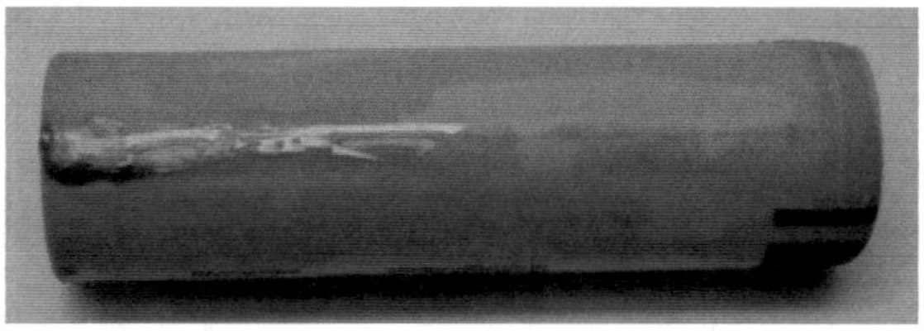

Cell had electrolyte leaks and an opened vent, but cells were shorted.

Near complete anode delamination due to current collector corrosion.

Anode delamination prevented electrode impedance measurements. Likely it was poor.

ejoseph.nemanick@aero.org EPL/Energy Technology Group

AEROSPACE

10

## Example reversed cell cathode

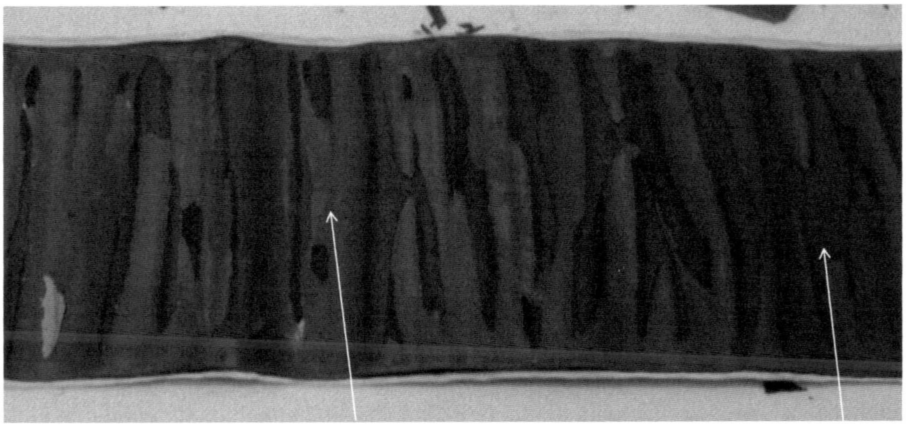

Cathode showed heavy copper plating

Delaminated area, where active material was adhered to separator

Heavy copper plating covered the cathode surface uniformly, and caused brittleness and adhesion to separator.

AEROSPACE

ejoseph.nemanick@aero.org EPL/Energy Technology Group

11

## Reversed cell cathode and separator

Separator with adhered cathode material

Cathode with heavy copper plating, with partially adhered separator.

Adhesion caused a separator split between layers.

Adhesion of the separator to the copper coated surface was severe enough to delaminate the separator.

AEROSPACE

ejoseph.nemanick@aero.org EPL/Energy Technology Group

12

## Separator damage, possible short areas

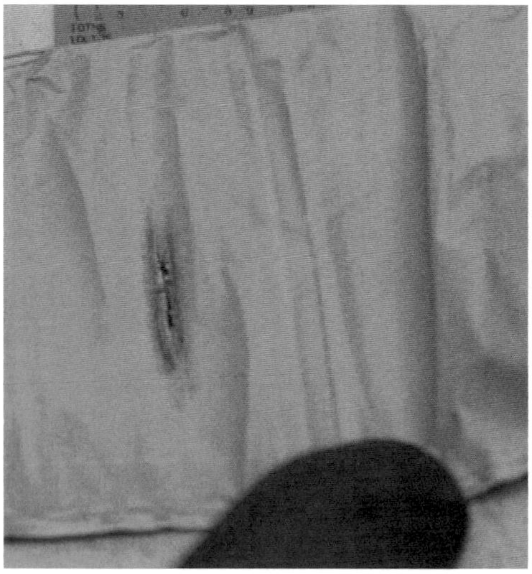

Damaged areas were brittle and darkened. Reverse side was dense, delaminated cathode material.

Each cell had several areas of discolored separator, typically in the center of the stack. Cathode material was heavily adhered in these areas. Likely short areas.



## Heavy corrosion of the copper current collector

### Cross sectional SEM images of anode samples

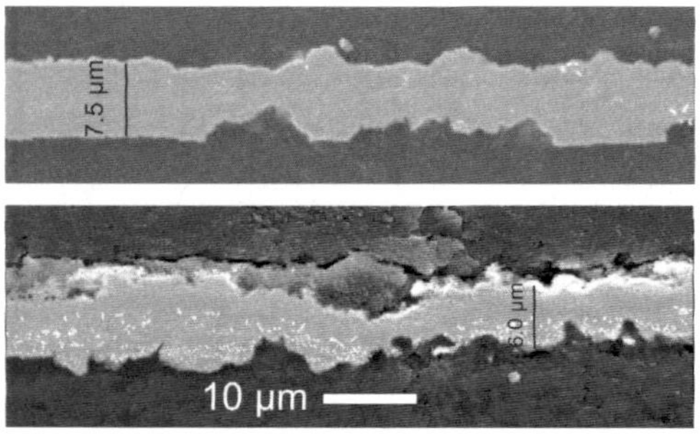


Reversed samples

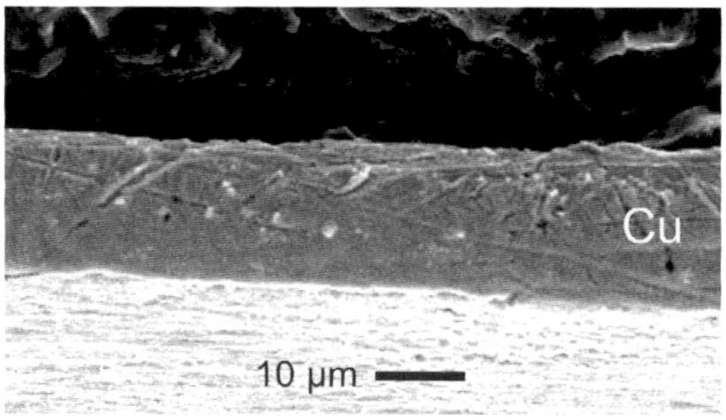


Control sample

The copper current collectors are heavily corroded in the reversed cells. Less than half of the original copper remains.

## Copper deposition on cathode surface

### Planar SEM images of cathode samples

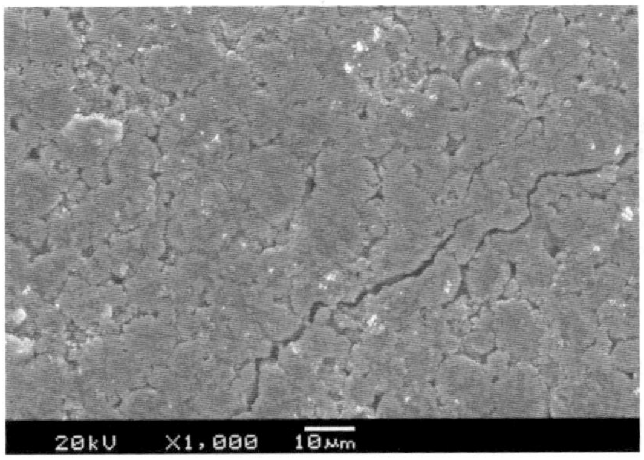


Reversed electrode samples

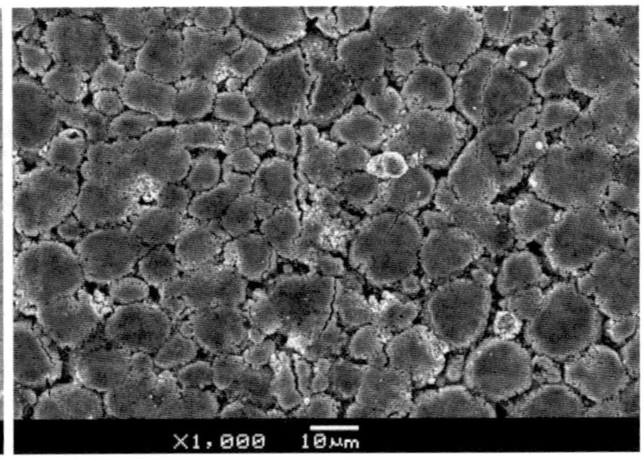


Control sample

Cathode surfaces of the reversed cells show densification of the surface of the electrode compared to the control. Copper has filled in the surface gaps.



## Dendritic copper deposition on cathode surface

### Cross sectional SEM images of cathode samples

Reversed cathode

Control sample

Copper dendrites

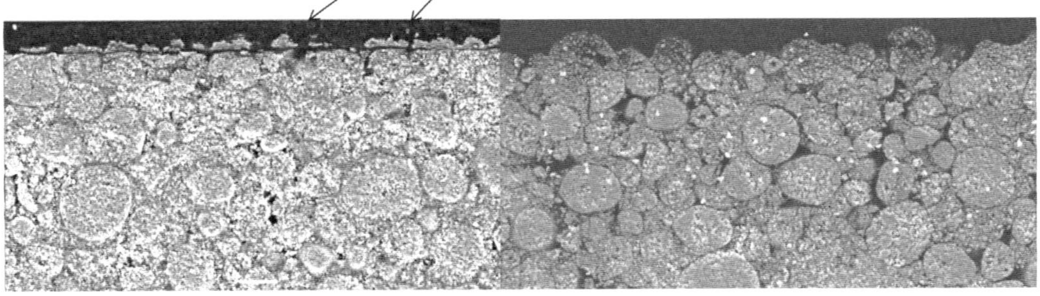

Copper dendrites were found on some but not all cathode surfaces. However, the separator had considerable areas of NCA adhesion, and dendrites may have been removed when cathode was pulled away from the separator.

Cathode surfaces of the reversed cells show densification of the surface of the electrode compared to the control. Copper has filled in the surface gaps.

## Intact copper dendrite examples on cathode surfaces

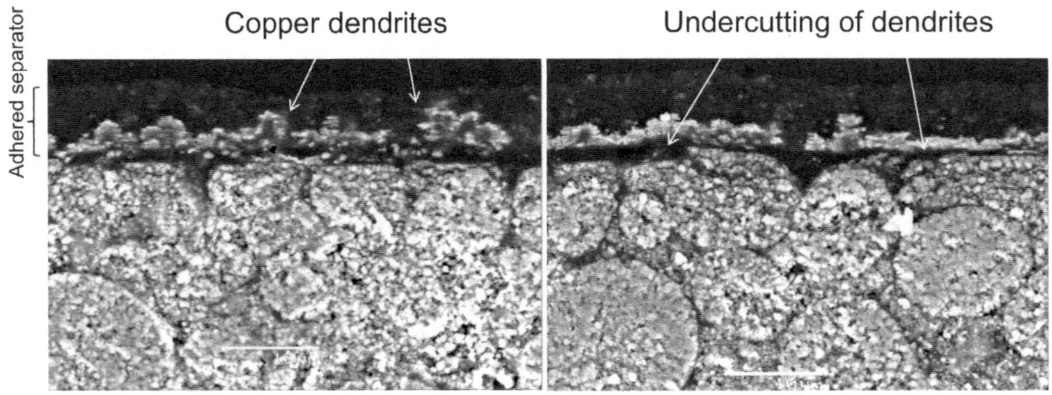


Undercutting of dendrites likely occurred during normal operation of cell, where the cathode potential was high enough to oxidize the base of dendrites, abandoning them in the separator.

Copper dendrites were preserved where the separator ceramic coating delaminated from the polymer portion of the separator.



## Copper penetration into the cathode material

Dendritic copper on surface

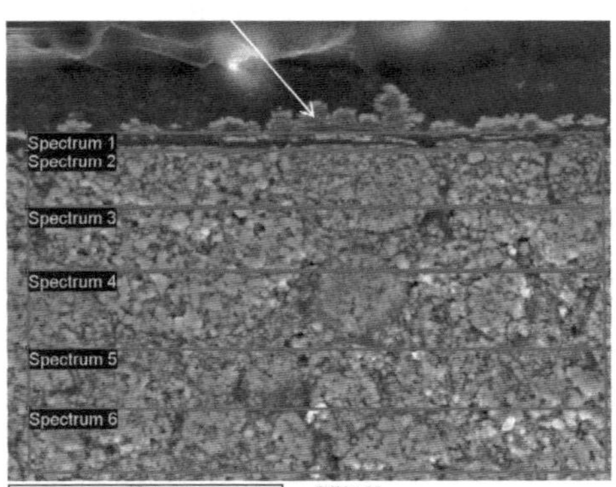


Depth of Copper (Cu/Ni ratio)

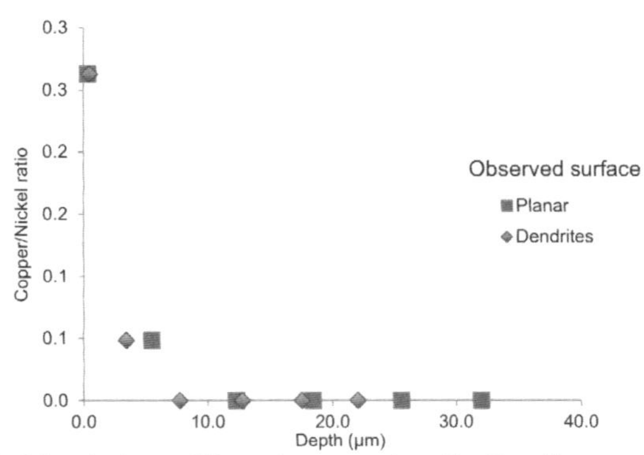


Samples without observed dendrites ('planar') did not show different copper localization than samples with dendrites on the surface.

Copper does not appear to grow from the Al current collector forming an all metal hard short. However, areas where the NCA is cracked may cause such a catastrophic short.

Copper deposition did not occur throughout the cathode. As expected, deposition was limited to the surface region of the electrodes.

## Copper dendrite growth through entire separator

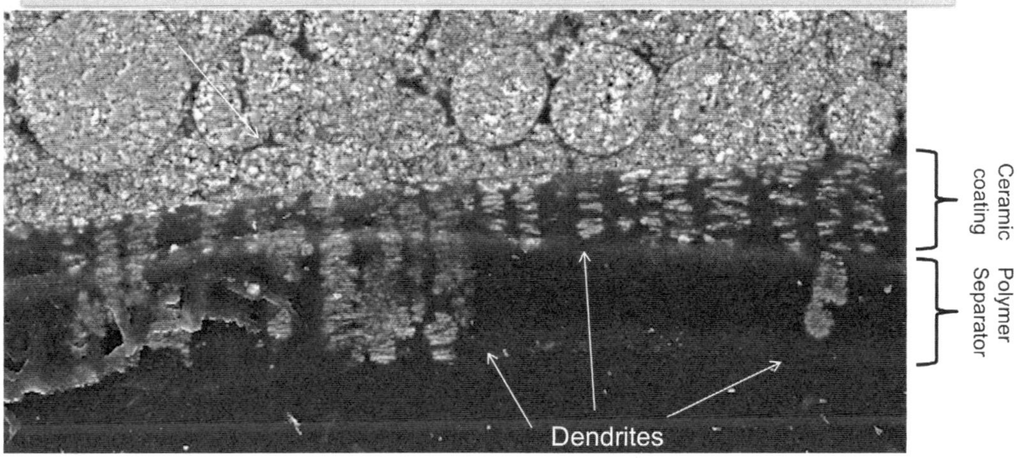


Samples of the damaged separator that were shown previously. These had large areas with dendrites that were through to the anode.

Dendrites were nearly uniformly found in the ceramic coating of the separator, but only certain areas show a high population of dendrites within the polymer portion.



## Copper dendrites in free separator

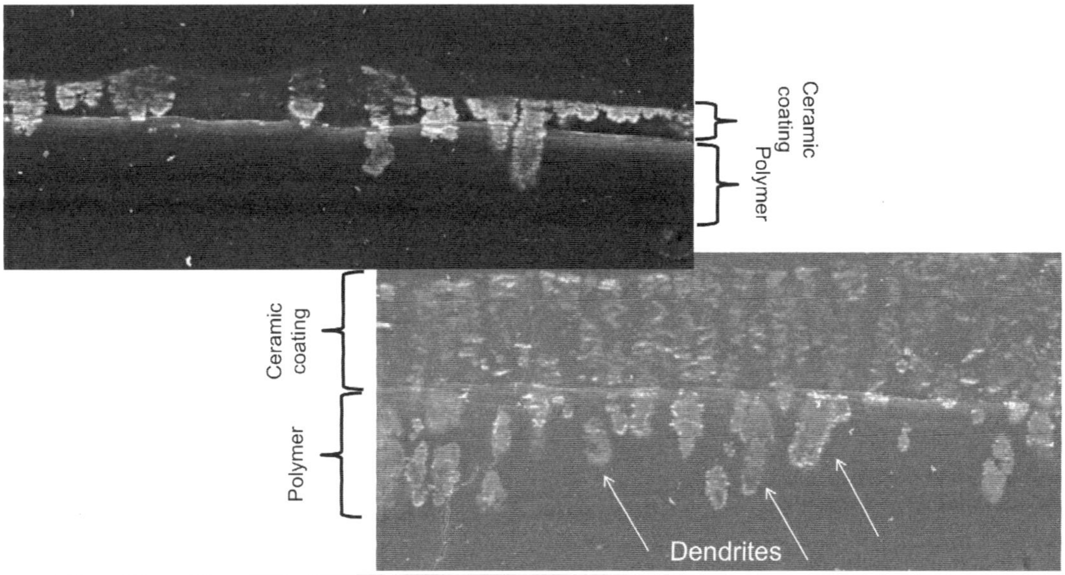


Dendrites did not show any locations where carbon was adhered to the dendrite filled separator. The short may have been a press contact.

## Conclusions

- Cells operated far into reversal, up to four cycles for 60% DOD GEO testing. This demonstrates a surprisingly high level of robustness for the cells. Cells did not go into thermal runaway when recharged, or immediately short.
- The shorting of one cell in the parallel pack did not prevent shorting of the other cells in the pack.
- Evidence of copper plating can be detected nondestructively using CT X-ray scans of the cell head spaces for sufficiently severe plating. This is the first nondestructive conclusive test for copper plating.
- Copper deposition was uniform and universal on the cathode surfaces, but not penetrating to the current collector.
  - Full dendrite penetration of the separator only occurred in a few places, shorting the cell and stopping further dendrite growth.



## Acknowledgements

- The author is grateful for the assistance and contributions of the following personnel at The Aerospace Corporation:
  - *Dr. Boyd Carter, Dr. Simon Liu, and Dr. Jeanne McGraw*
- The Aerospace Corporation is gratefully acknowledged for supporting this work as part of its Sustained Experimentation and Research for Program Applications (SERPA).

## 40% DOD - 99% recharge--Voltages

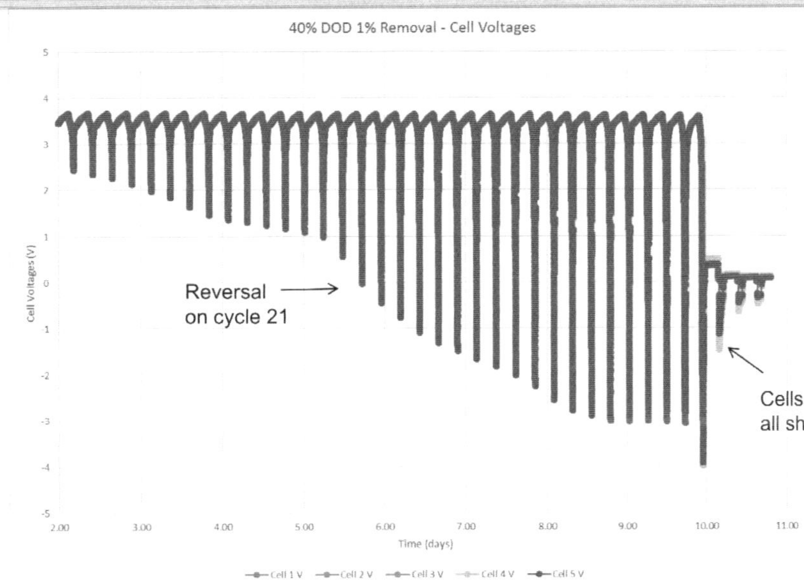


Cells go into reversal after 21 cycles, but operate for 17 additional cycles with discharge partially in reversal.



## 40% DOD - 99% recharge--Temperatures

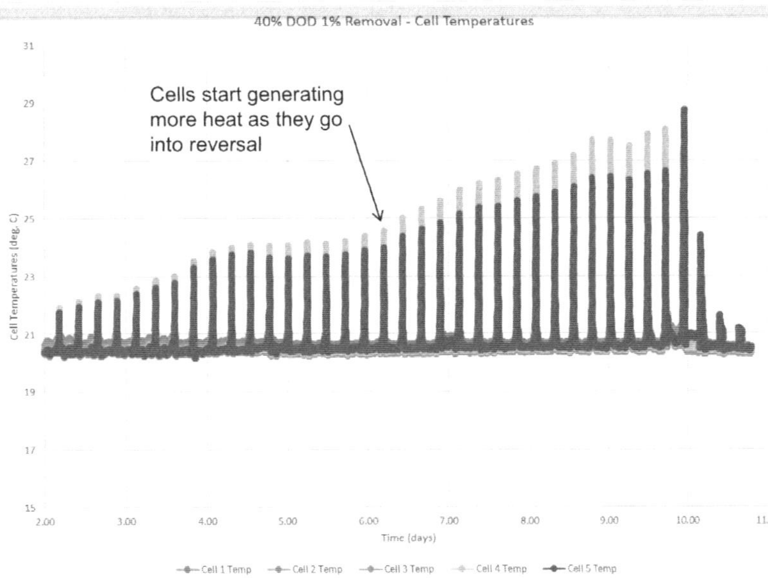


Cells show an increasing temperature as they head below 0% SOC and into reversal. Cell temperature drops after all the cells fail short.

## 40% DOD - 99% recharge--Currents

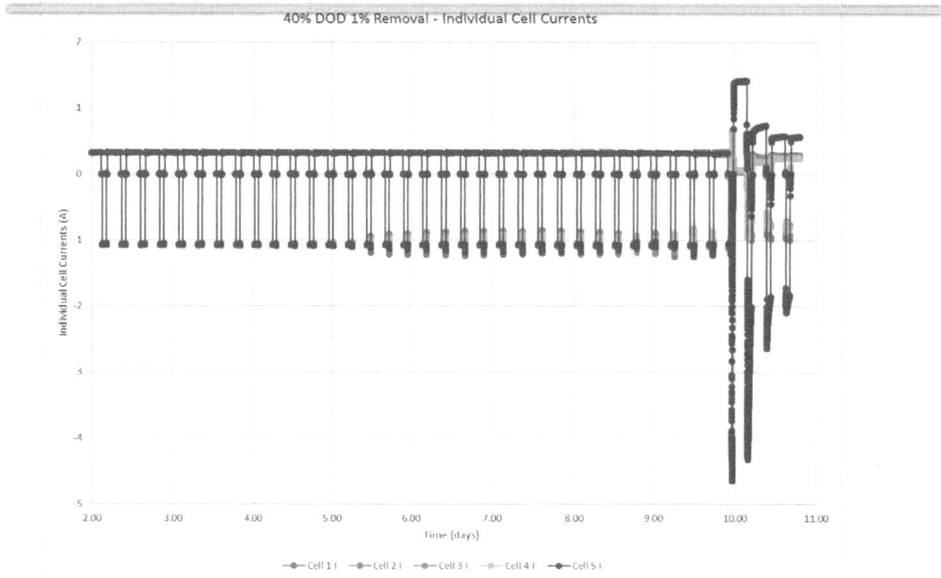


After cells go into reversal, current sharing is not equal. Cell 5 appears to fails first, and takes the majority of the current before the other cells short.



## 40% DOD - 99% recharge--Conclusions

- Cells operated far into reversal, with 19 full cycles from the start of reversal to first cell shorting.
- Cells in reversal did not share current equally. Cell 4 appears to have had a higher impedance in reversal as shown by increased temperature and decreased currents.
- The initial short occurred in cell 5, as shown by current sharing. However, the other cells quickly shorted as well, creating a more equal current sharing.

# Battery Safety Requires More than Safe Batteries

**2016 International Battery Seminar**

**Todd Vanyo – Systems and Safety Manager**



## What is a safe system?

**Let's make the supposition that the batteries themselves are safe. What techniques exist to keep them in a safe operating area?**

## Well Known Solutions – Protector

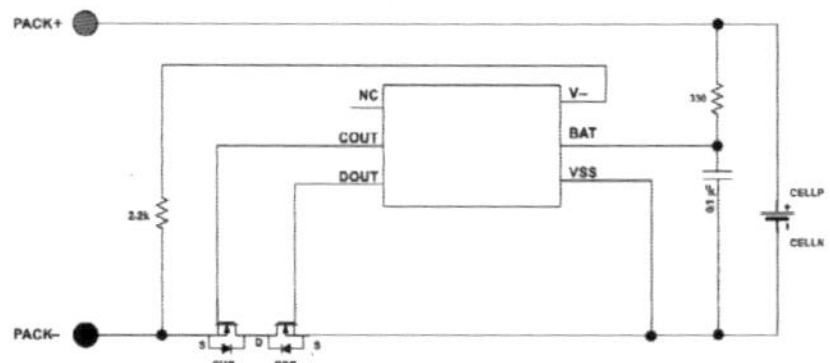


**Simple voltage protectors turn off the charge / discharge path when the battery reaches appropriate threshold**

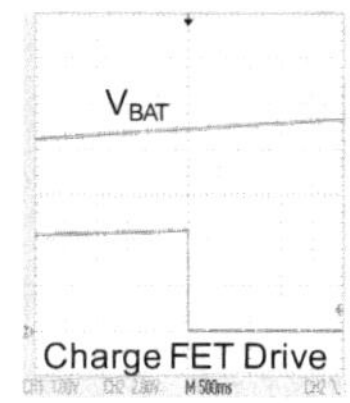


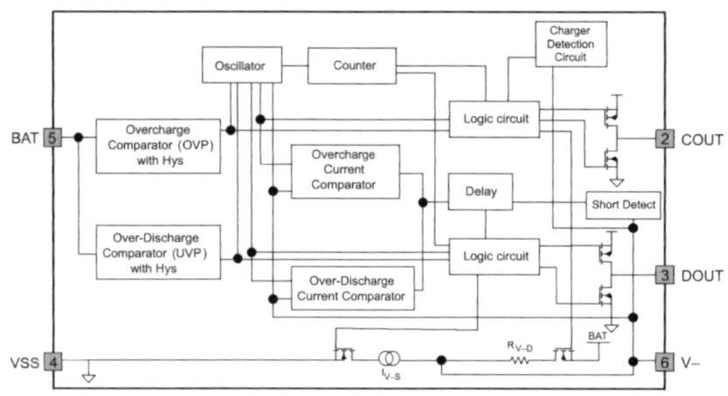


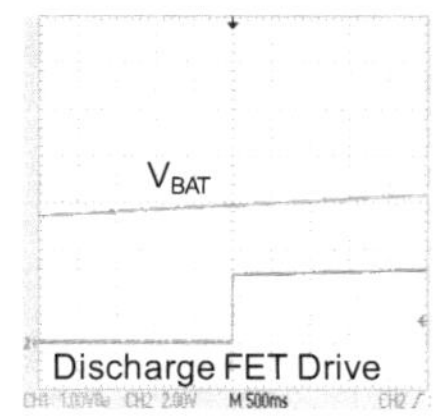


## Well Known Solutions – Charger

**A "smart" charger:**

- **Monitors battery voltage**
- **Limits Input Current**
- **Monitors Temperature**
  - **Limits voltage & current per JEITA**
- **Timers to disable charging**
- **Monitors internal charger currents**

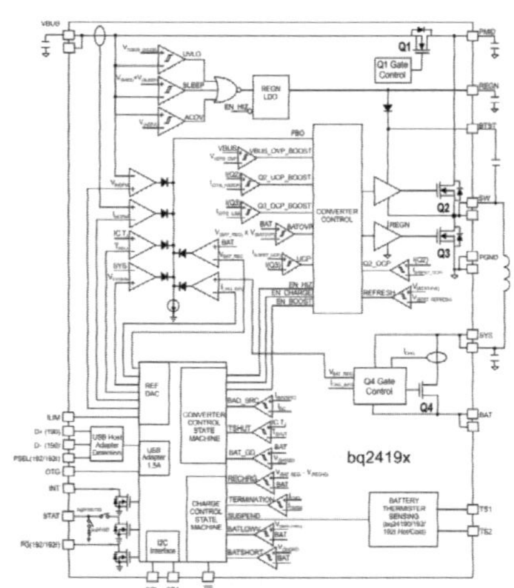


**Is there another approach to facilitate safety?**

## The Typical Catalog Device

In "catalog" products, comparator thresholds are specified, designed, characterized, and then tested at the factory to check whether they are functioning to the spec.

| | | | | | | |
|---|---|---|---|---|---|---|
| $V_{SC\_SYS}$ | SYS Short circuit (SC) detection threshold | BAT FET OFF if $V_{SYS} < V_{SC\_SYS}$ | 1.60 | 1.80 | 2.00 | V |
| | SYS SC detection hysteresis | | 25 | 50 | | mV |
| $I_{BATOC}$ | Battery Overcurrent | | 3.1 | | 6.2 | A |

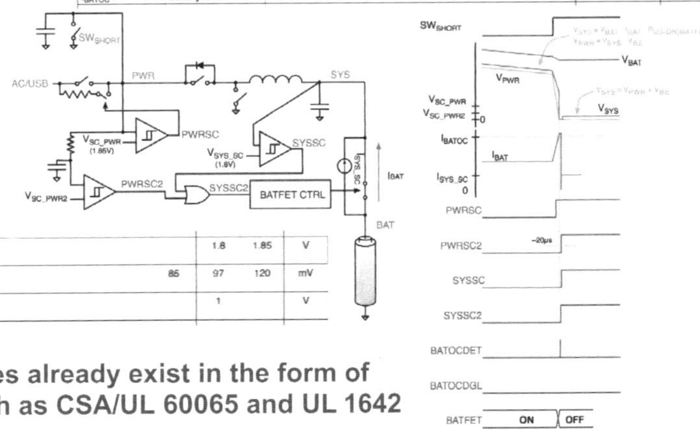


| | | | | | | |
|---|---|---|---|---|---|---|
| $V_{SC\_PWR}$ | PWR soft short detection threshold | $V_{PWR}$ decreasing | | 1.8 | 1.85 | V |
| | PWR short hysteresis | $V_{PWR}$ increasing | 85 | 97 | 120 | mV |
| $V_{SC\_PWR2}$ | PWR hard short detection threshold | $V_{PWR}$ decreasing | | 1 | | V |

**A number of approaches already exist in the form of prescriptive specs such as CSA/UL 60065 and UL 1642**

## So...

**...assuming that good design practices are followed and qualified and acceptable products are used, is there anything else that can be done?**

## A more radical suggestion has been made…

"[Venkat] Viswanathan has a recommendation for companies making vape pens and other gadgets that use lithium-ion batteries: **Crib from automakers making electric cars.**"

http://www.wired.com/2016/02/exploding-e-cigs-and-vape-pens/

## How Does The Automotive Industry Do It?

INTERNATIONAL STANDARD

ISO 26262-1

First edition 2011-11-15

Road vehicles — Functional safety —

Certain semiconductor products are subject to a stringent set of requirements designed to enable diagnosis of device failure modes in order to meet the safety goal…

**ISO 26262-1 falls into a category called hazard based standards, requiring analysis during the design cycle of the IC.**

## Safety Failures and their Causes

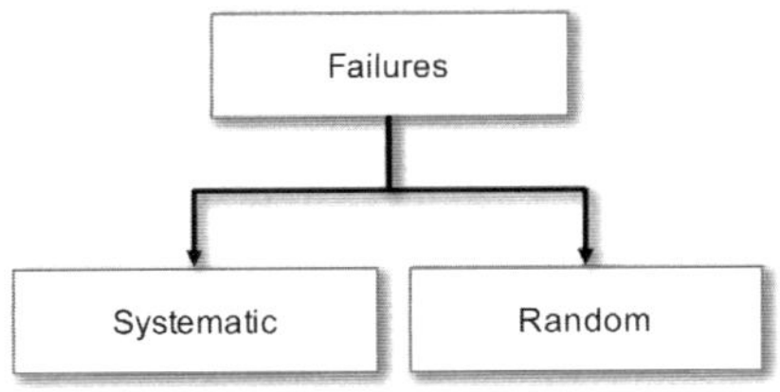


Systematic:

- Result from a failure in design or manufacturing; result of failure to follow best practices
- Rate of systematic failures can be reduced through continual and rigorous process improvement

Random:

- Result from random defects inherent to process or usage condition; generally cannot be reduced.
- Focus must be on the detection and handling of random failures in the application.

**Organizational *culture* impacts occurrences of systematic failures!**

**Implementing safety mechanisms can change impact of random failures!**

## Safety Mechanisms Reduce Risk

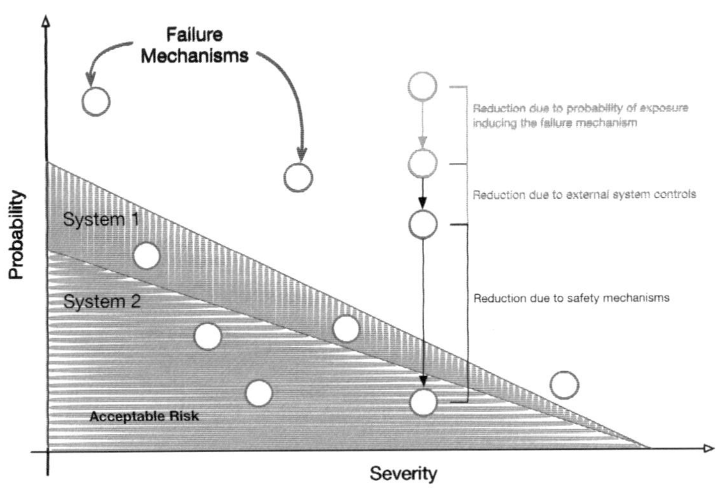


**Goal is to decrease risks to an acceptable level, in light of application specifics**

## Safety Mechanisms Examples At A Glance

How is it known that, in operation, the short circuit protections are working?

How is it known that the comparator thresholds are still valid?

| | | | | | | |
|---|---|---|---|---|---|---|
| $V_{SC\_SYS}$ | SYS Short circuit (SC) detection threshold | BAT FET OFF if $V_{SYS} < V_{SC\_SYS}$ | 1.60 | 1.80 | 2.00 | V |
| | SYS SC detection hysteresis | | 25 | 50 | | mV |
| $I_{BATOC}$ | Battery Overcurrent | | 3.1 | | 6.2 | A |
| $V_{SC\_PWR}$ | PWR soft short detection threshold | $V_{PWR}$ decreasing | | 1.8 | 1.85 | V |
| | PWR short hysteresis | $V_{PWR}$ increasing | 85 | 97 | 120 | mV |
| $V_{SC\_PWR2}$ | PWR hard short detection threshold | $V_{PWR}$ decreasing | | 1 | | V |

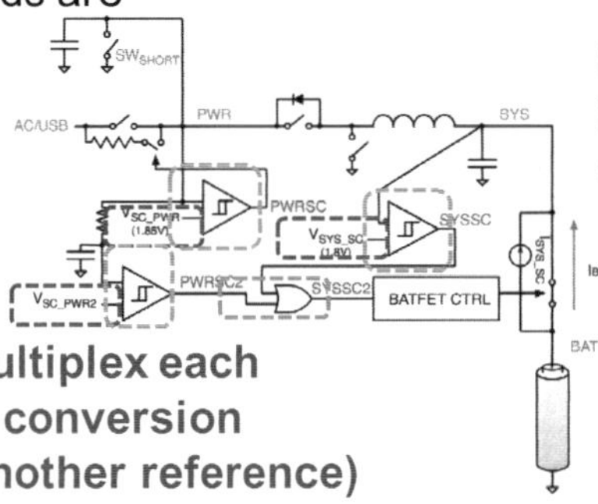


Force comparator inputs to a voltage to cause trip

Force logic levels to observe expected changes in the logic chain

**Add an ADC and multiplex each reference into it for conversion (means you need another reference)**

## Who Controls It?

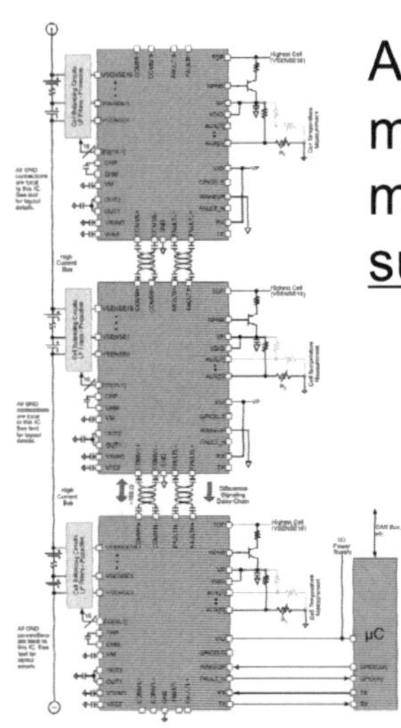


Automotive systems usually have a micro-controller or host processor to manage the process. <u>This device is also subject to safety requirements!</u>

Stand-alone devices need more decision making capability, which increases silicon area and complexity.

## Metrics for Risk Reduction

Evaluates level of diagnostic coverage and safe faults vs. undetected faults

Based on safety goal ASIL

$$SPFM = 1 - \frac{\Sigma \lambda_{DU}}{\Sigma \lambda}$$

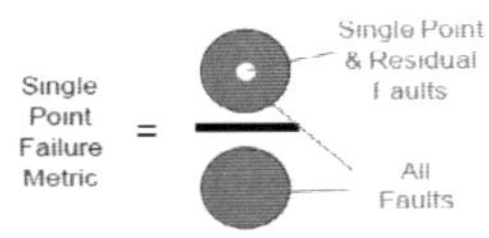


Based on Failure Rates of Faults that may lead to a violation of a safety goal

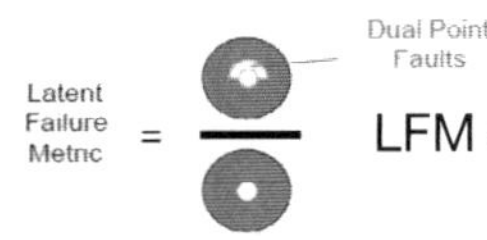


$$LFM = 1 - \frac{\Sigma LAT_{undetected}}{\Sigma \lambda - \Sigma \lambda_{DU}}$$

"Failure Diagnosis and Prognosis for Automotive Systems", by *Tom Fuhrman, General Motors R&D, IFIP Workshop, June 25-27, 2010*

| Metric | ASIL B | ASIL C | ASIL D |
|---|---|---|---|
| Single Point | > 90% | > 97% | > 99% |
| Latent Fault | > 60% | > 80% | > 90% |

**For xEV battery systems, the majority of the market requires ASIL C performance from the battery monitor**

## In Conclusion...

- The market must determine what level of risk is acceptable
  - Are generally accepted engineering practices, in the form of simple and cost effective protection mechanisms already called out by prescriptive standards like CSA/UL 60065 acceptable?
- The techniques exist to make battery systems safer
  - Hazard based standards require additional effort by IC vendors, pack makers, and system designers to properly implement these techniques
  - They do not come for free

ENERGY AND TRANSPORT SCIENCES LABORATORY
Department of Mechanical Engineering, Texas A&M University

# Mesoscale Implications in Lithium-Ion Battery Thermal Behavior

**Chien-Fan Chen, Ankit Verma, and Partha P. Mukherjee**

*Energy and Transport Sciences Laboratory, Mechanical Engineering Department, Texas A&M University, College Station, TX 77843*

## Outline

- Objective and Background
- Microstructure-aware Electrochemical-Thermal Model
- Influence of Electrode Microstructure on Thermal Behavior
- Relation between Cell Capacity and Thermal Behavior
- Conclusion

# Lithium-ion Battery

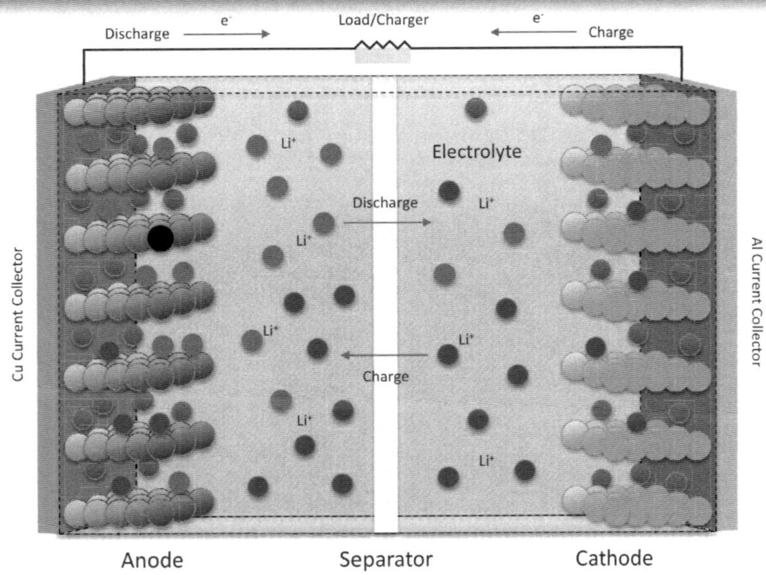


Schematic diagram of a lithium ion battery showing the different domains usually considered while modeling an electrode.



# Objective

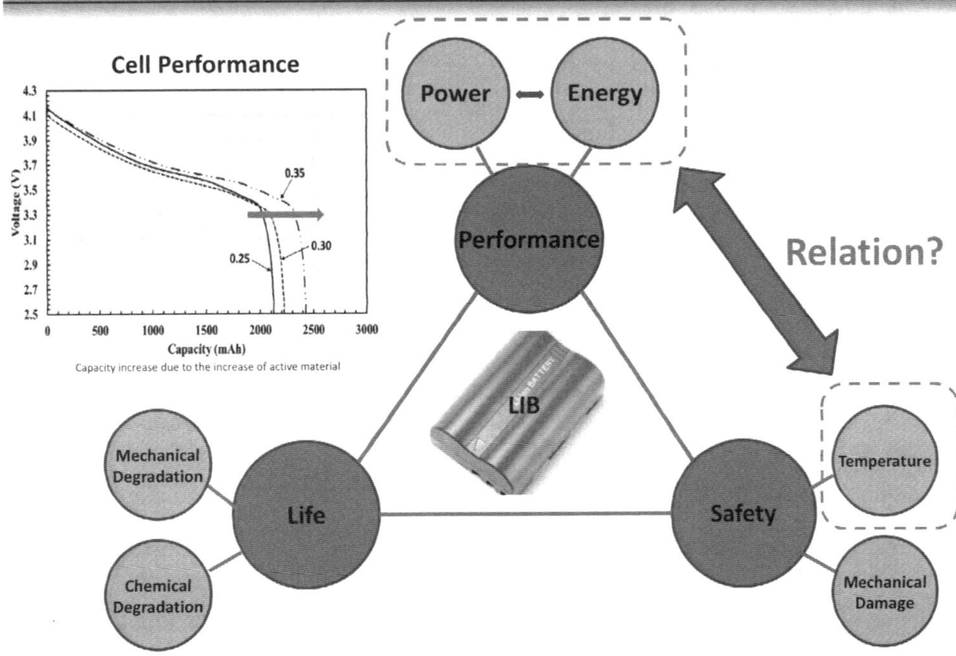

# Electrode Microstructure

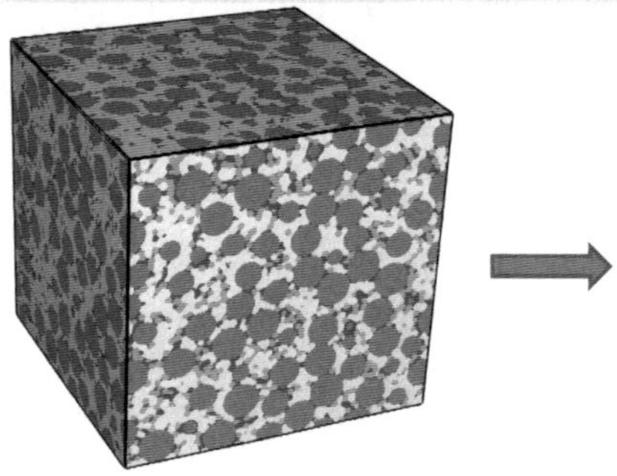

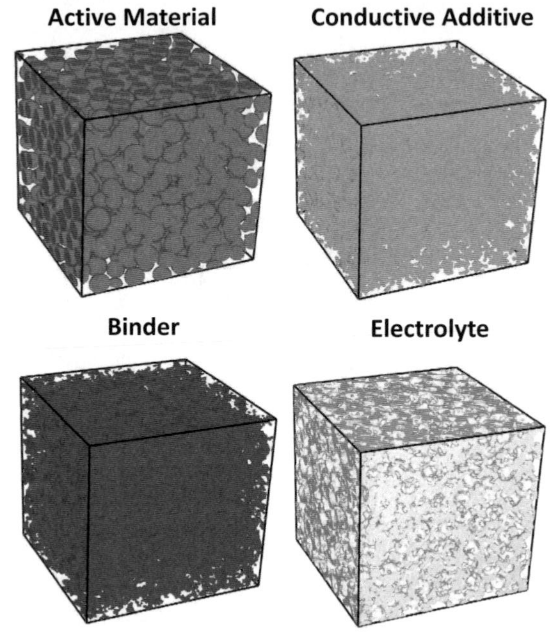


- Active material
  - Electrochemical reaction
  - Store Lithium ion
- Conductive Additive
  - Increase electrical conductivity
- Binder
  - Bind active materials and conductive additives
- Electrolyte
  - Transport lithium ion



# Microstructure Dependent Thermal Behavior

**AM:** 62 vol.% **Porosity:** 20 vol. %

**AM:** 54 vol. % **Porosity:** 30 vol. %

**AM :** 46 vol. % **Porosity:** 40 vol. %

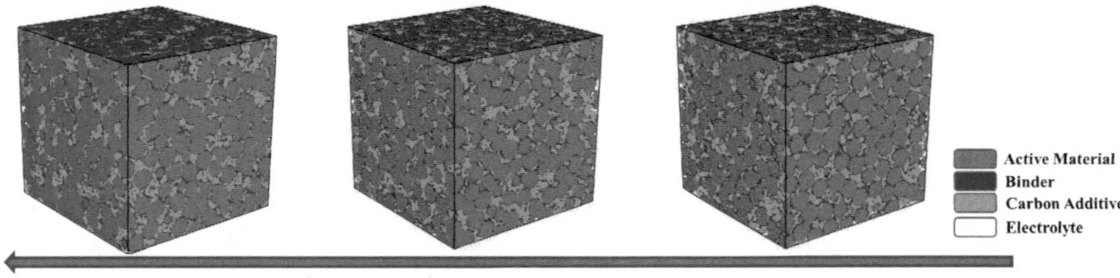


**Theoretical Capacity Increase**

N/P Loading Ratio → Open-circuit potential → **Reversible Heat** $jT(dU/dT)$

Porosity, Tortuosity → Ion concentration and potential gradient → **Joule Heat** $\sigma_s \nabla\phi_s \nabla\phi_s + \kappa\nabla\phi_e\nabla\phi_e + \kappa_D \nabla \ln c_e \nabla\phi_e$

Surface Area → Overpotential → **Reaction Heat** $j(\phi_s - \phi_e - U)$

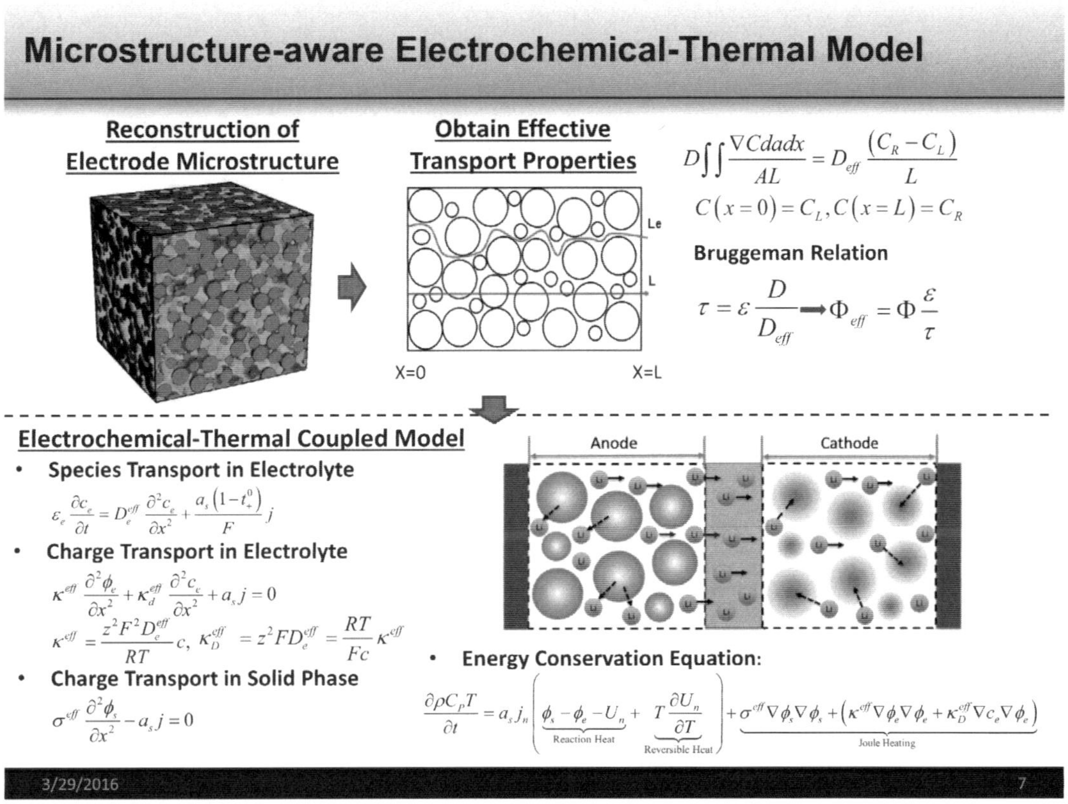
Microstructure-aware Electrochemical-Thermal Model
Reconstruction of Electrode Microstructure
Obtain Effective Transport Properties
Le
L
X=0
X=L
$D\iint \frac{\nabla C dadx}{AL} = D_{eff}\frac{(C_R - C_L)}{L}$
$C(x=0)=C_L, C(x=L)=C_R$
Bruggeman Relation
$\tau = \varepsilon \frac{D}{D_{eff}} \rightarrow \Phi_{eff} = \Phi \frac{\varepsilon}{\tau}$
Electrochemical-Thermal Coupled Model
Species Transport in Electrolyte
$\varepsilon_e \frac{\partial c_e}{\partial t} = D_e^{eff}\frac{\partial^2 c_e}{\partial x^2} + \frac{a_s(1-t_+^0)}{F} j$
Charge Transport in Electrolyte
$\kappa^{eff}\frac{\partial^2 \phi_e}{\partial x^2} + \kappa_d^{eff}\frac{\partial^2 c_e}{\partial x^2} + a_s j = 0$
$\kappa^{eff} = \frac{z^2F^2D_e^{eff}}{RT}c,\ \kappa_D^{eff} = z^2FD_e^{eff} = \frac{RT}{Fc}\kappa^{eff}$
Charge Transport in Solid Phase
$\sigma^{eff}\frac{\partial^2 \phi_s}{\partial x^2} - a_s j = 0$
Anode
Cathode
Energy Conservation Equation:
$\frac{\partial \rho C_p T}{\partial t} = a_s j_n\left(\phi_s - \phi_e - U_n + T\frac{\partial U_n}{\partial T}\right) + \sigma^{eff}\nabla\phi_s\nabla\phi_s + \left(\kappa^{eff}\nabla\phi_e\nabla\phi_e + \kappa_D^{eff}\nabla c_e\nabla\phi_e\right)$
Reaction Heat
Reversible Heat
Joule Heating
3/29/2016
7

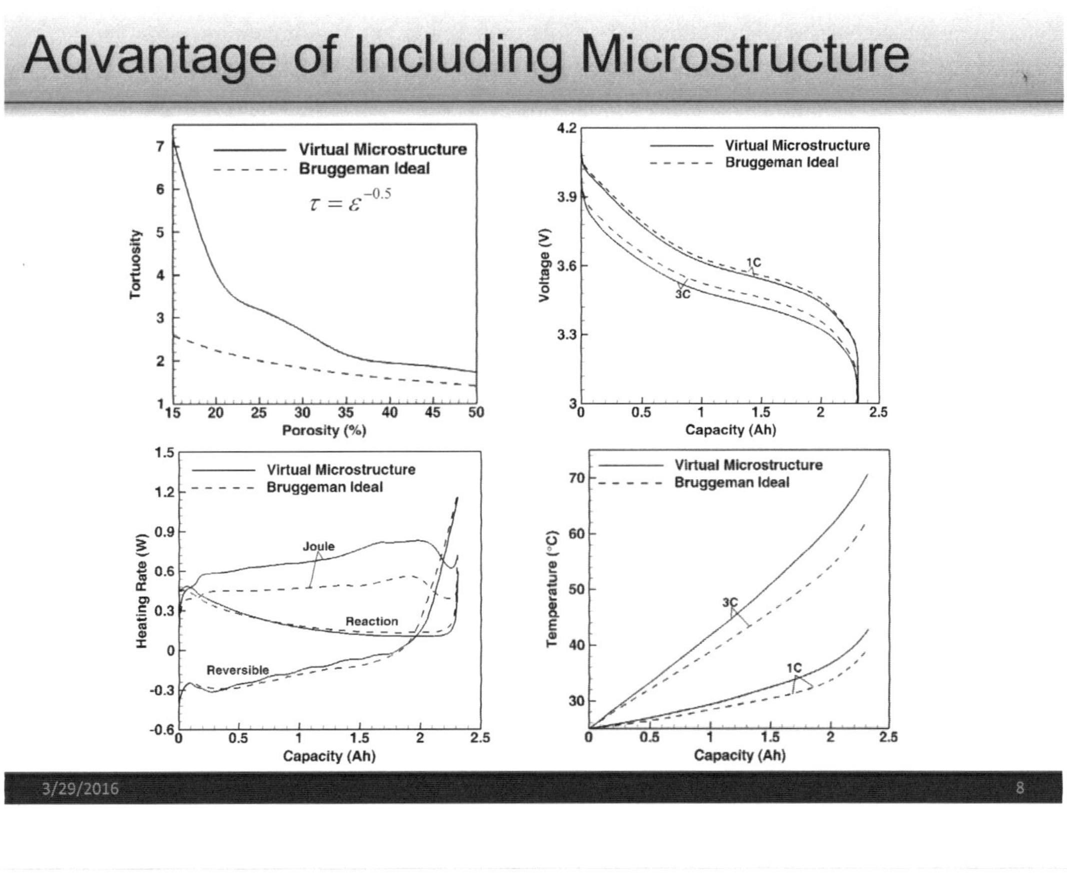
Advantage of Including Microstructure
Virtual Microstructure
Bruggeman Ideal
$\tau = \varepsilon^{-0.5}$
Tortuosity
Porosity (%)
Voltage (V)
Capacity (Ah)
1C
3C
Heating Rate (W)
Joule
Reaction
Reversible
Temperature (°C)
3/29/2016
8

# Influence of C-rate

$$\frac{\partial \rho C_P T}{\partial t} = a_s j_n \left( \underbrace{\phi_s - \phi_e - U_n}_{\text{Reaction Heat}} + \underbrace{T\frac{\partial U_n}{\partial T}}_{\text{Reversible Heat}} \right) + \underbrace{\sigma^{eff}\nabla\phi_s\nabla\phi_s + \left(\kappa^{eff}\nabla\phi_e\nabla\phi_e + \kappa_D^{eff}\nabla c_e\nabla\phi_e\right)}_{\text{Joule Heating}}$$

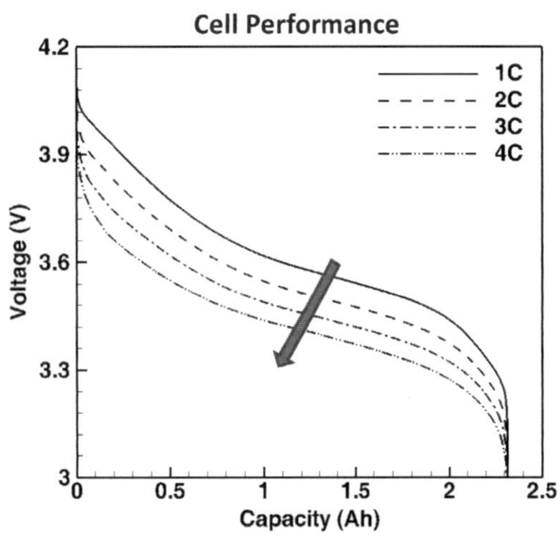


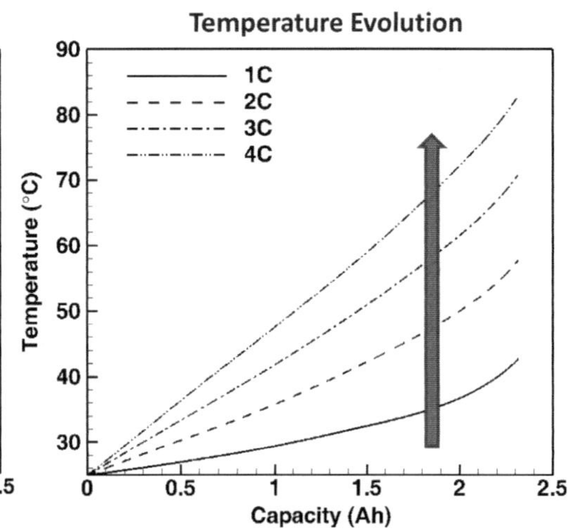




# Heat Generation Inside the Electrode

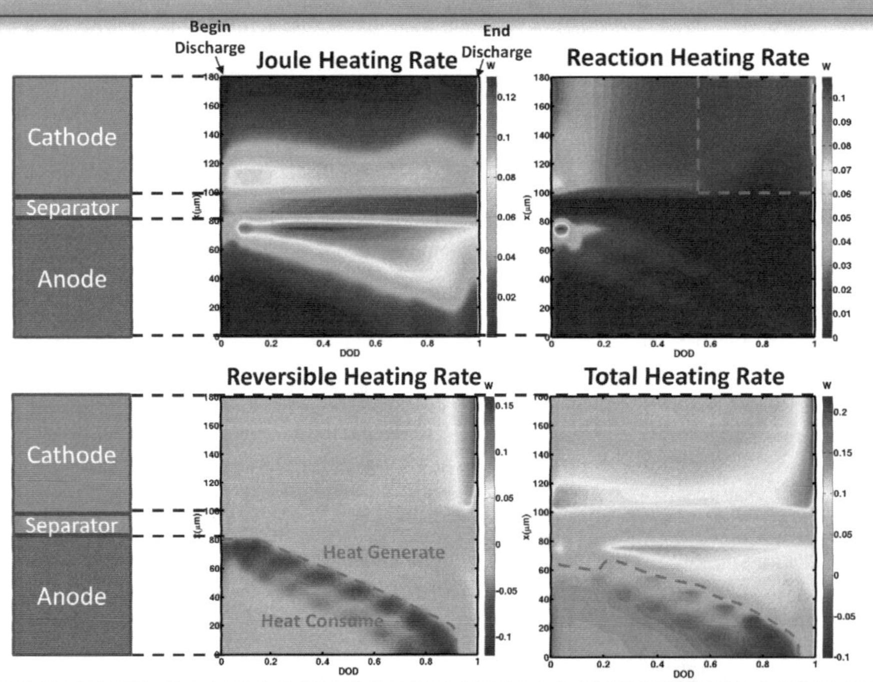

# Influence of Microstructure

**Anode:** AM→ 95 wt.%, Porosity → 0.25
**Output Current:** 6A
**Operation Condition:** Adiabatic, $T_{amb}$=25°C

*Volume Fraction of active Material and corresponding porosity*

| Porosity, ε | 0.15 | 0.20 | 0.25 | 0.30 | 0.35 | 0.40 |
|---|---|---|---|---|---|---|
| AM vol.% | 66 | 62 | 58 | 54 | 50 | 46 |

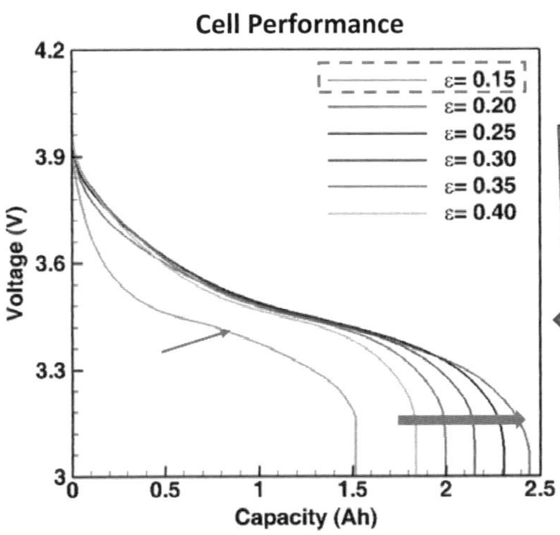


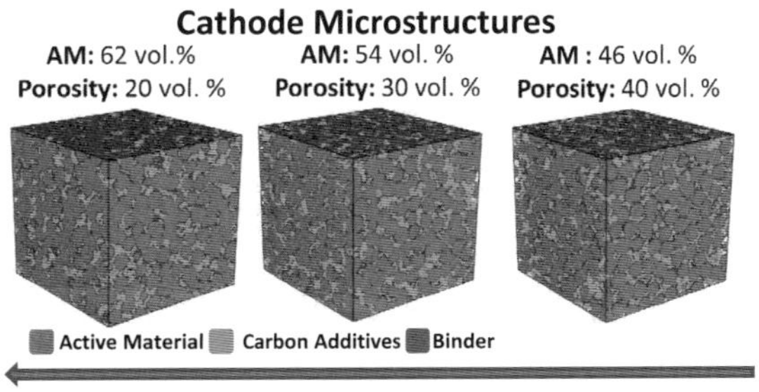


- Increase theoretical capacity
- Decrease porosity
- Increase tortuosity
- Increase internal resistance

3/29/2016 11

# Influence of Microstructure

**Anode:** AM→ 95 wt.%, Porosity → 0.25
**Output Current:** 6A
**Operation Condition:** Adiabatic, $T_{amb}$=25°C

*Volume Fraction of active Material and corresponding porosity*

| Porosity, ε | 0.15 | 0.20 | 0.25 | 0.30 | 0.35 | 0.40 |
|---|---|---|---|---|---|---|
| AM vol.% | 66 | 62 | 58 | 54 | 50 | 46 |

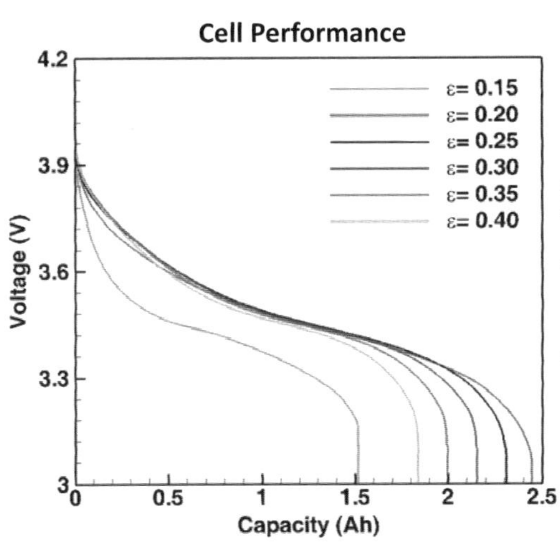


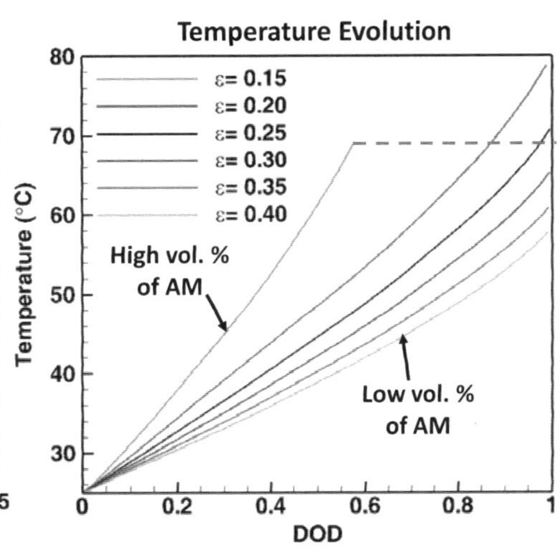


3/29/2016 12

# Influence of Microstructure (2)

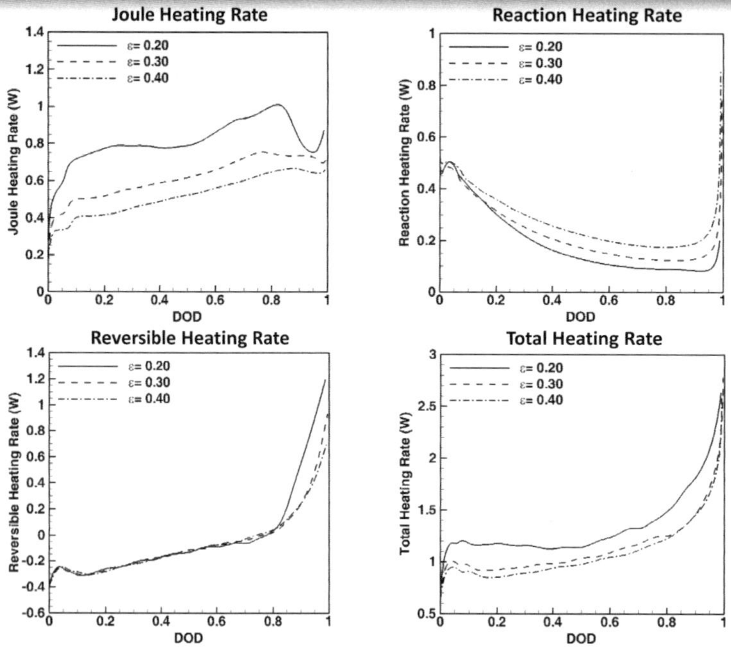


3/29/2016 13

# Joule Heating Rate

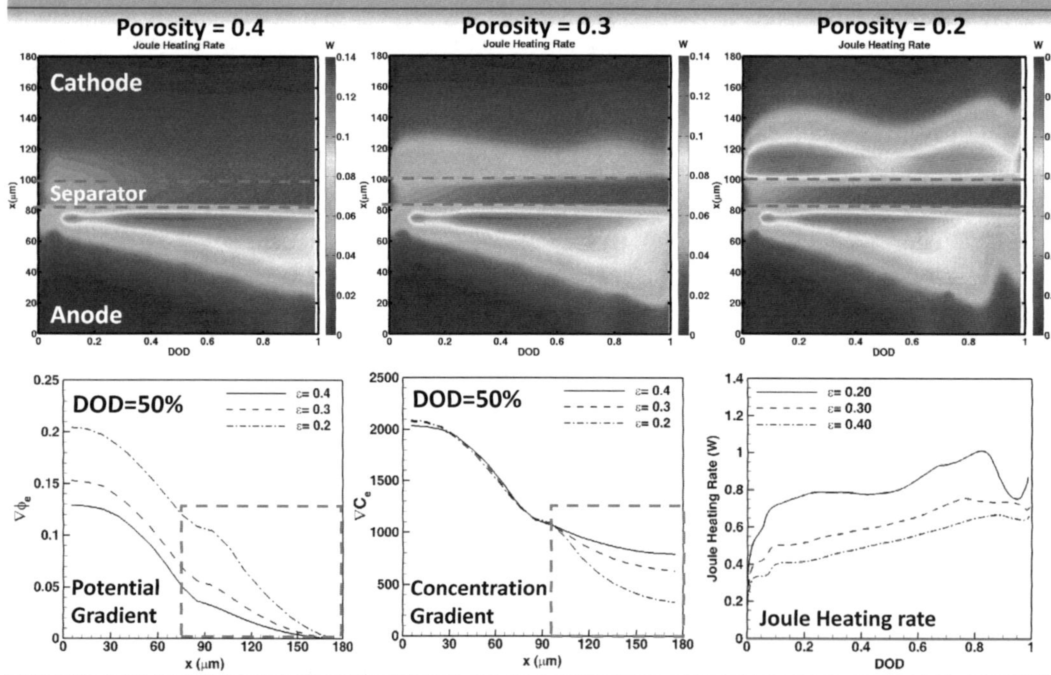


3/29/2016 14

# Reaction Heating Rate

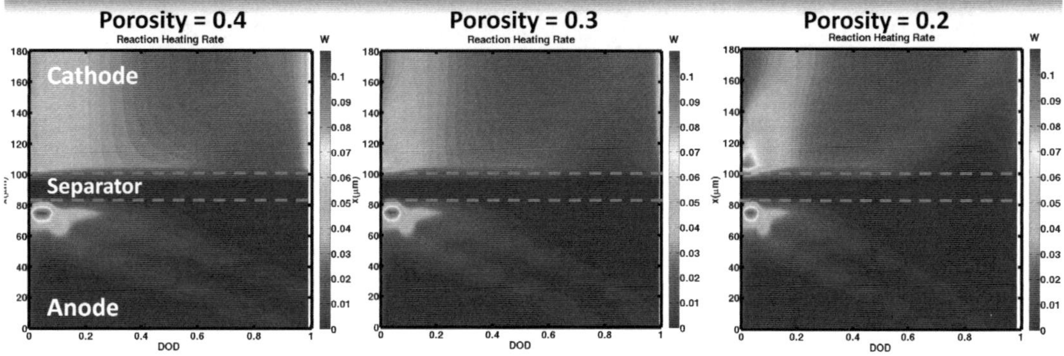


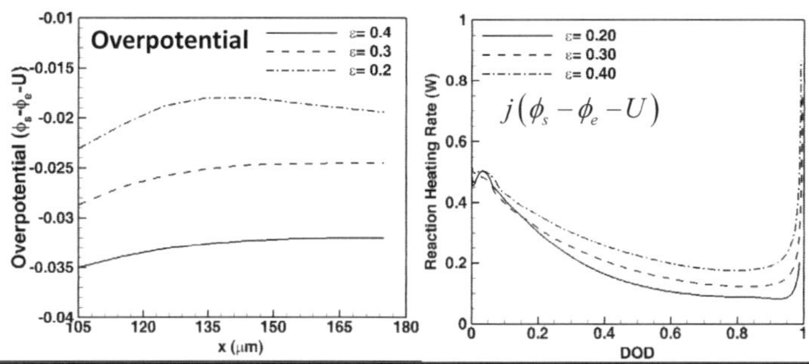


- The overpotential is affect my the active interfacial area between electrolyte and active material.



# Reversible Heating Rate

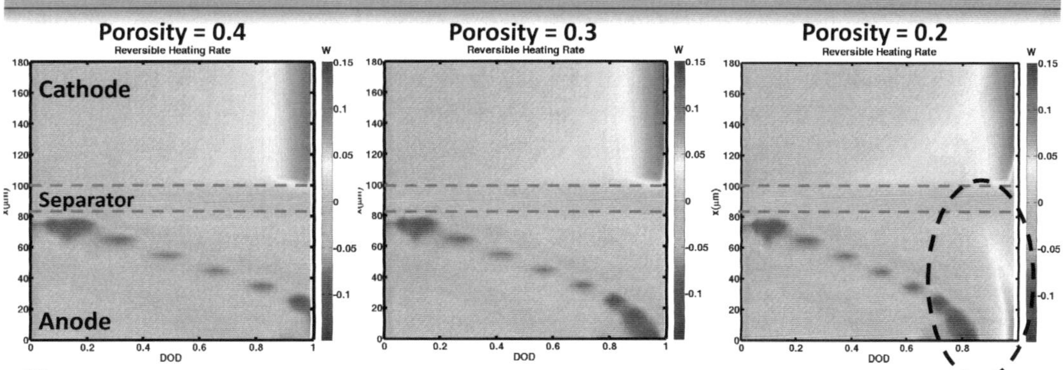


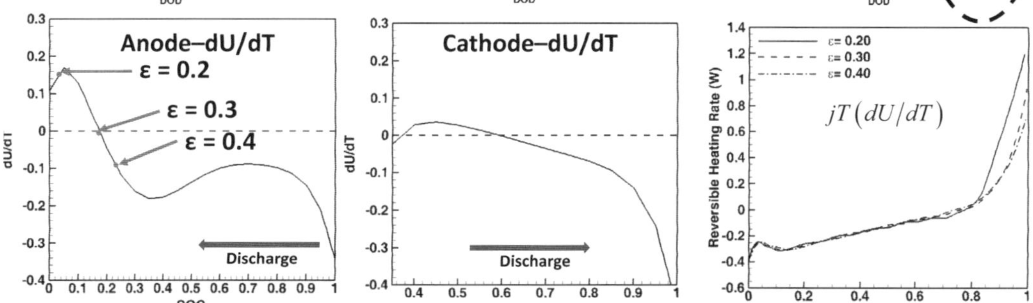


 K. E. Thomas and J. Newman, *Journal of Power Sources*, vol. 119, pp. 844-849, Jun 1 2003. 

# Total Heating Rate

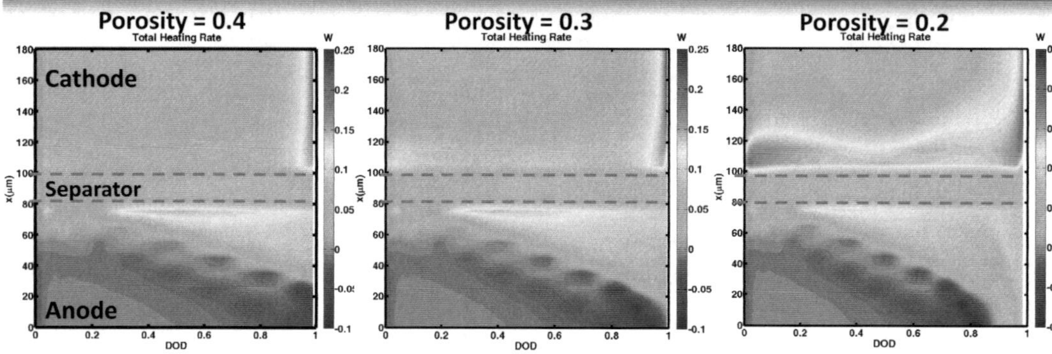


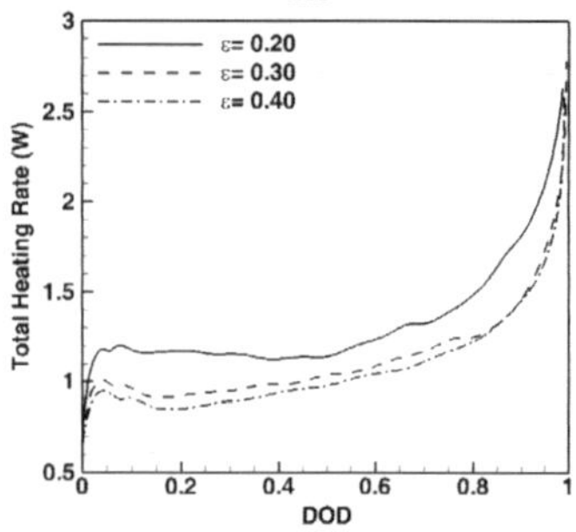


- Heat generation from anode is mainly from ohmic heat.
- Heat generation is large
  (a) in the end of discharge.
  (b) Near the separator
- The N/P loading ratio has influence on the heat generation when the loading (Ah/cm$^2$) of anode and cathode is close.



# Safety Design Window

- **Unsafe**: T > 80°C, thermal runaway
- **Potential Risk**: 80°C >T > 60°C, increase degradation
- **Safe:** 60°C >T, safe operation temperature

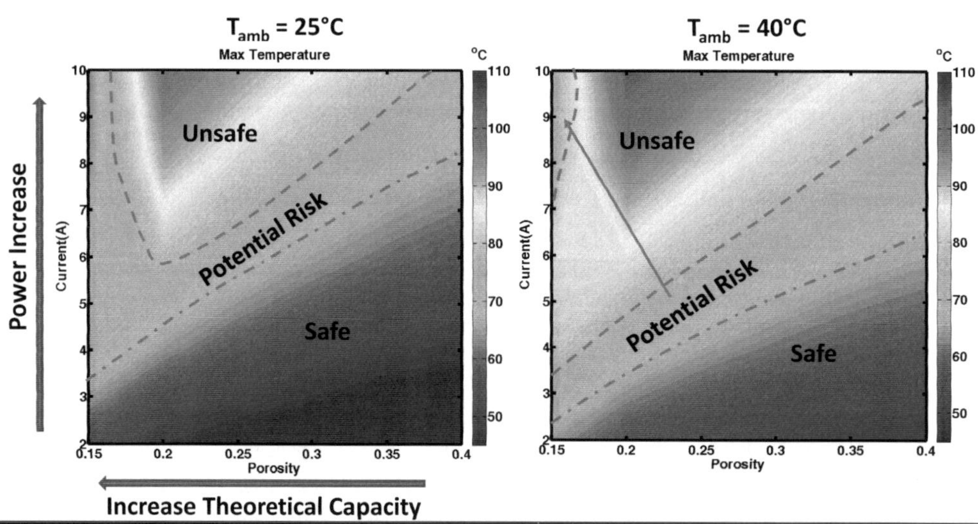

# Capacity Variation due to External Cut-off

- To avoid thermal runaway, the battery can be shut down with external control, and Positive temperature coefficient (PTC).
- Capacity lost can happen since the cell cannot be operate with full capacity.

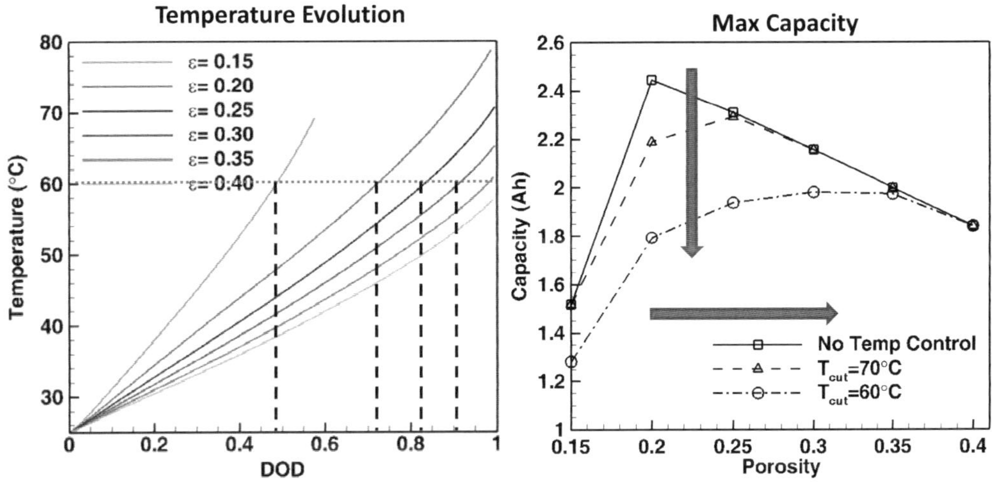




# Temperature Control with Microstructure

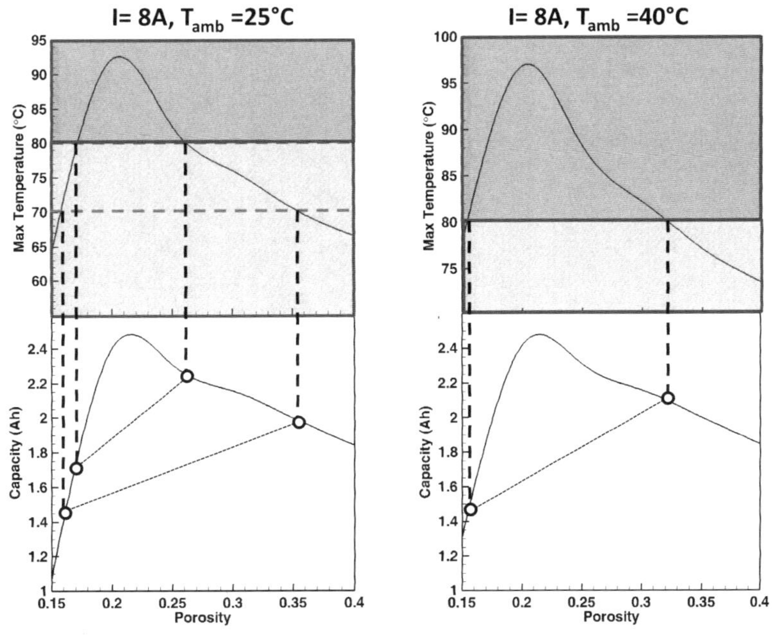

# Temperature Control with Microstructure

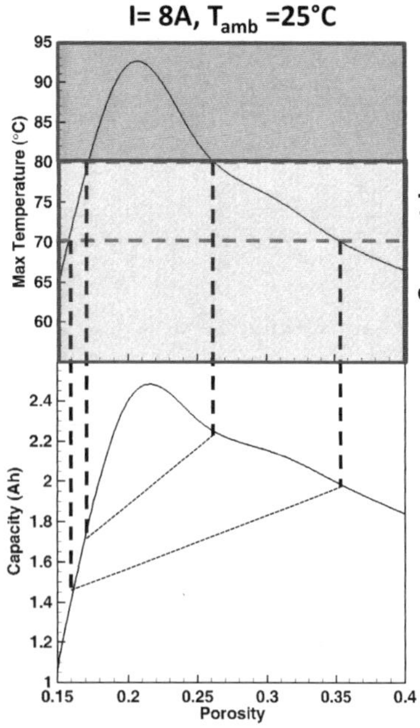


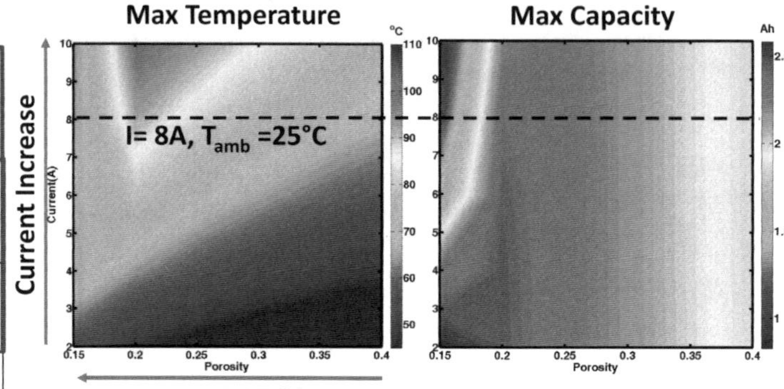


1. Active Material Increase
2. Porosity Decrease

- **Output high current (power) is controlled:**

→ Low porosity electrode



# Temperature Control with Microstructure

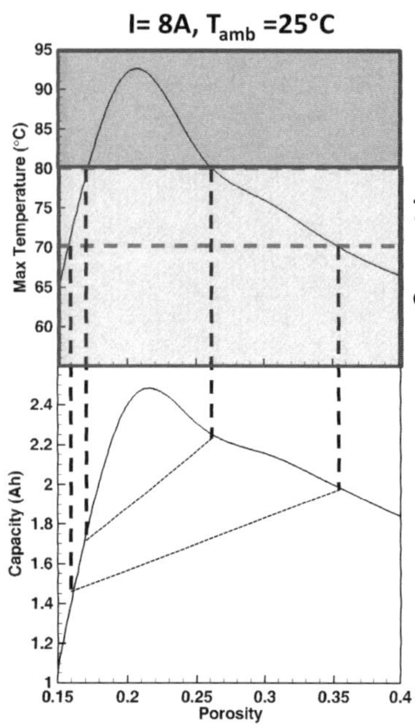


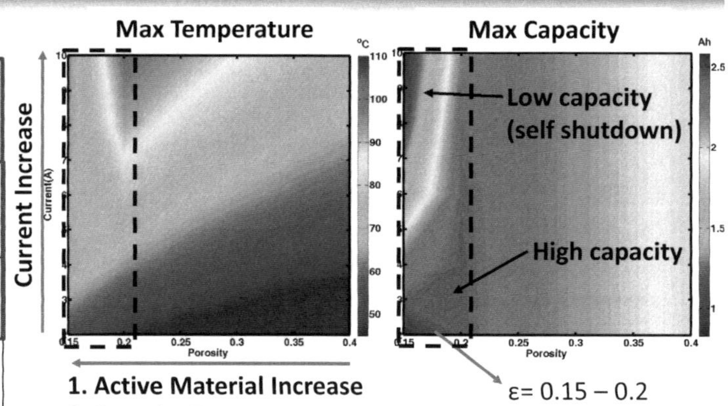


1. Active Material Increase
2. Porosity Decrease

- **Output high current (power) is controlled:**

→ Low porosity electrode

- **Output high current (power) is unintentional (ex. Extern short):**

→ High porosity electrode (ε < 0.2)

# Influence of Electrode Thickness

**Anode:** AM→ 95 wt.%, Porosity → 0.25
**Cathode:** AM → 90 wt.%, Porosity → 0.25
**Output Current:** 6A
**Operation Condition:** Adiabatic, $T_{amb}$=25°C

**Cell Performance**

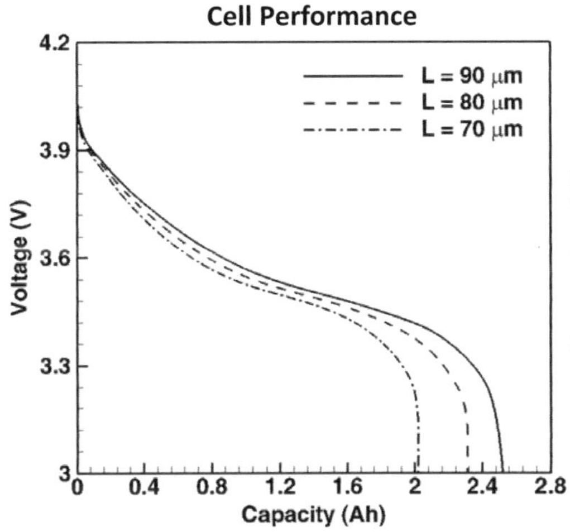


**Temperature Evolution**

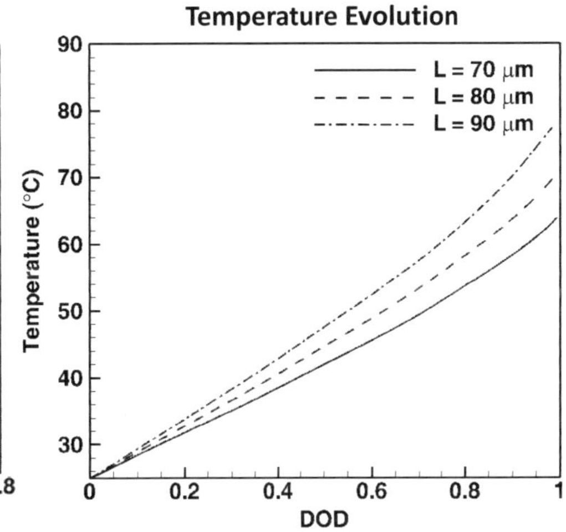




# Influence of Electrode Thickness (2)

**Joule Heating Rate**

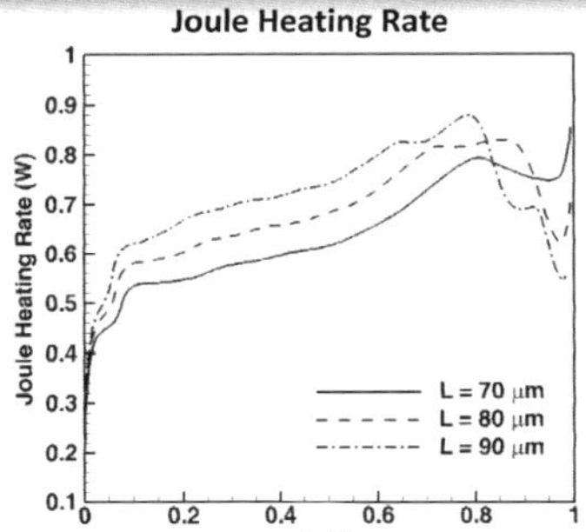


**Reaction Heating Rate**

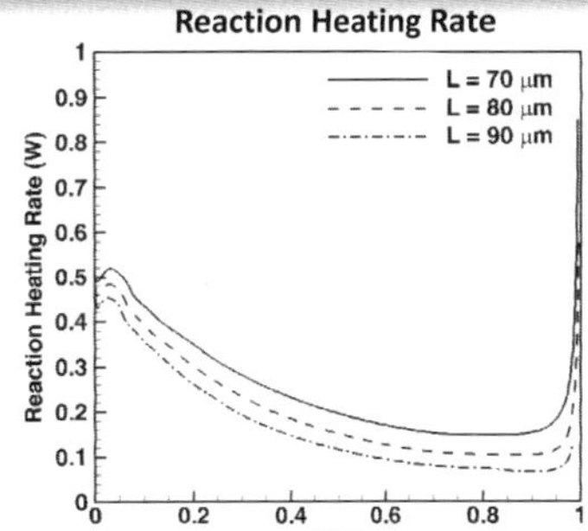


**Reversible Heating Rate**

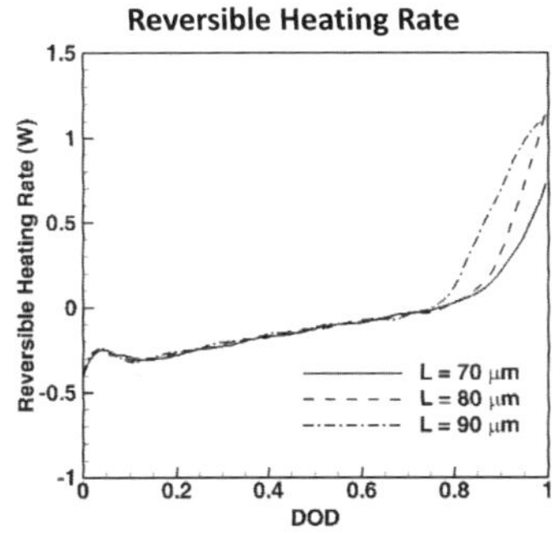


**Total Heating Rate**

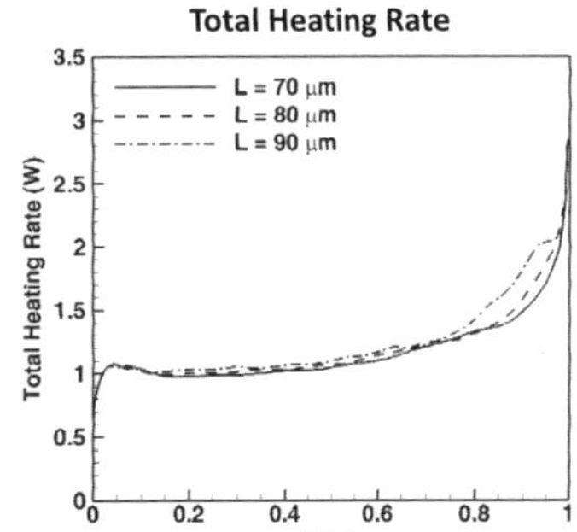

# Heating Rate

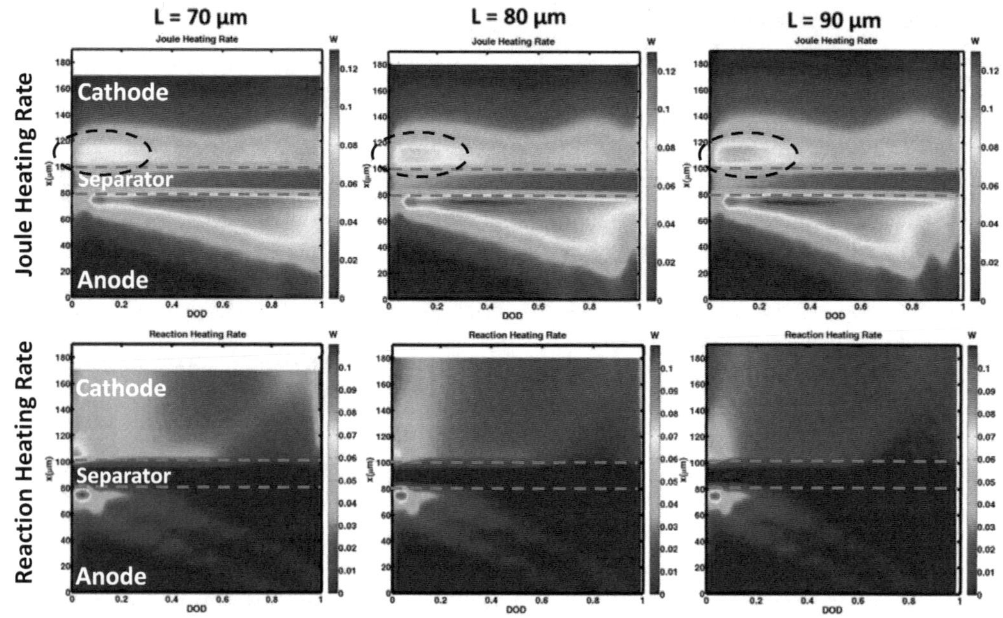


3/29/2016 25

# Heating Rate (2)

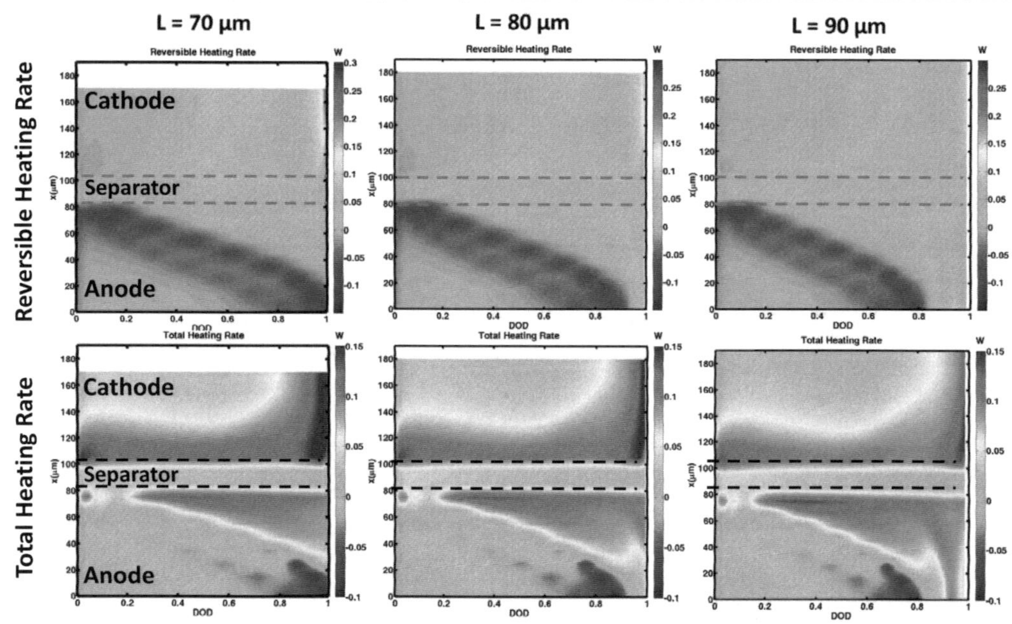


3/29/2016 26

# Decrease Porosity or Increase Thickness?

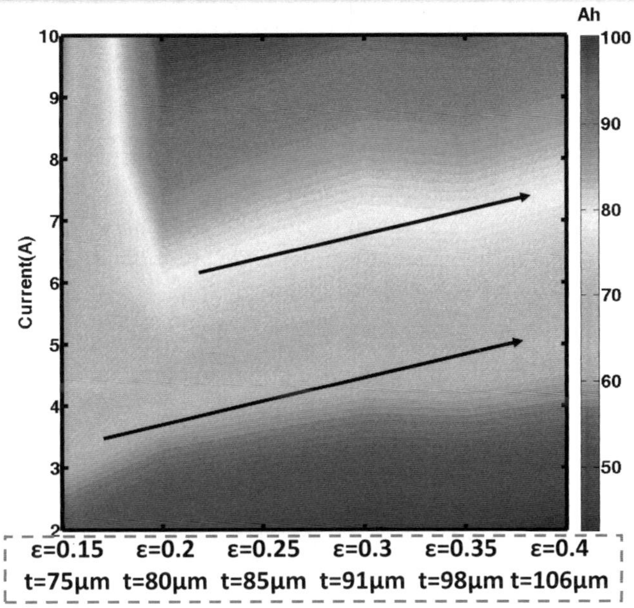


**Operation Condition:**
Adiabatic, $T_{amb}$=25°C
**Theoretical Capacity:**
2.476 Ah

- To maintain the battery safety and cell capacity, increase electrode thickens is a battery way to increase the capacity of cell.

➡ **Theoretical Capacity = 2.476 Ah**



# Microstructure and Low Temperature Operation

**Anode:** AM→ 95 wt.%, Porosity → 0.25
**Output Current:** 4A
**Operation Condition:** Adiabatic, $T_{amb}$=-10°C

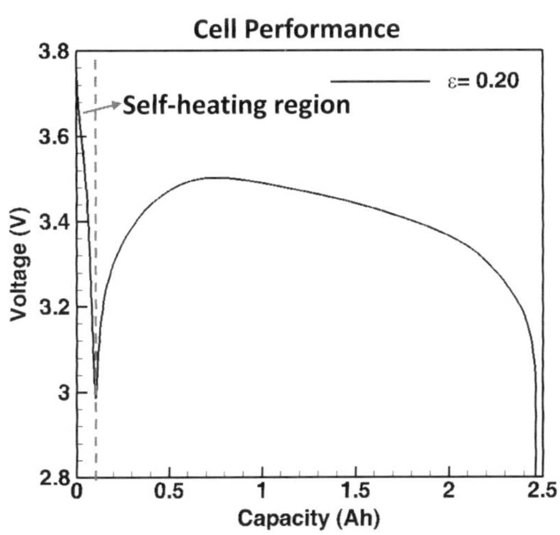


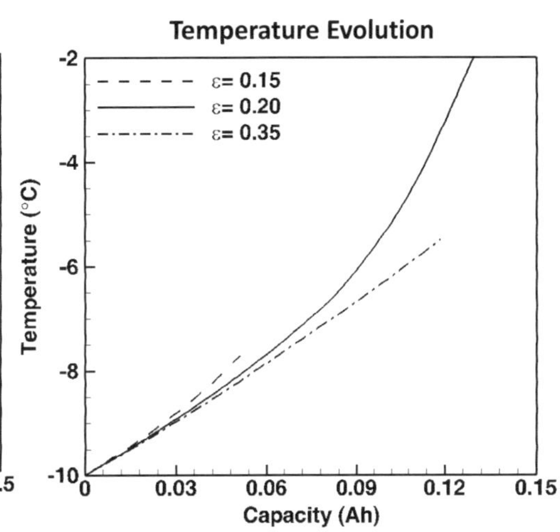

# Conclusion

- Electrode microstructure has influence on the thermal behavior of cell.

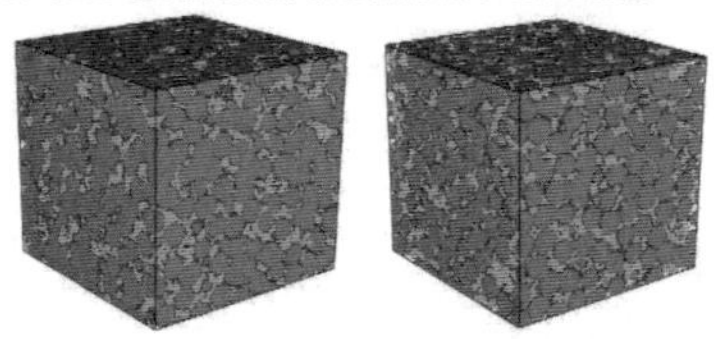

**Increase AM→ Increase Capacity**

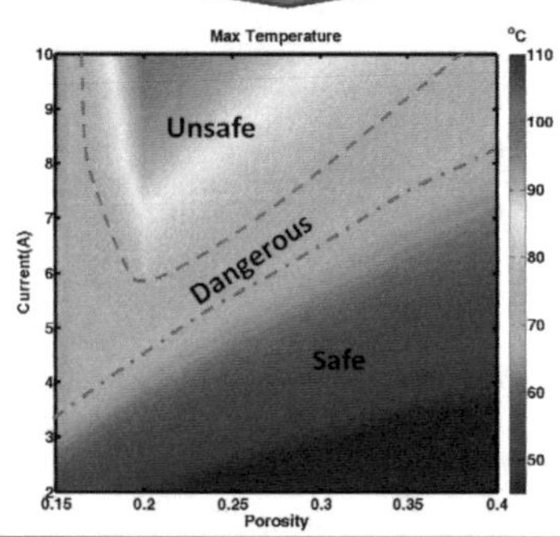


- With thermal control, the capacity of cell can be compromised.
- Microstructure dependent thermal behavior can be used to control the temperature of cell.
- To increase capacity, the method of increasing electrode thickness is better than decreasing electrode porosity with the concept of safety concern.

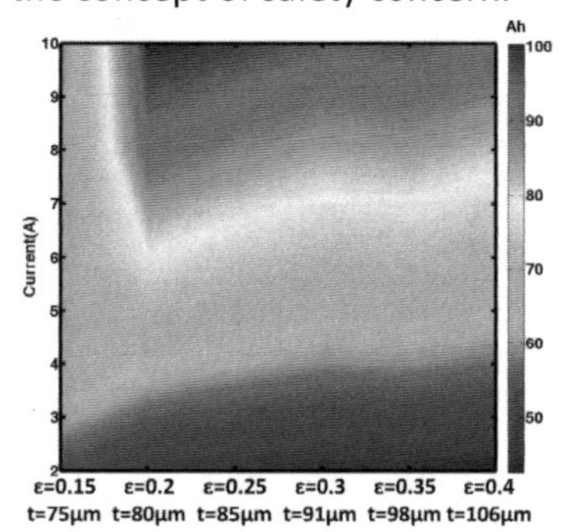




# Acknowledgement

Financial support from DOE EERE CAEBAT (Computer Aided Engineering for Batteries) program is gratefully acknowledged.

# Thank You!

**International Battery Seminar**
250 First Avenue, Suite 300
Needham, MA 02494
USA

ISBN 978-1-5108-2254-2

CURRAN ASSOCIATES INC.
proceedings.com